W9-AUB-702

PERSPECTIVES
ON OUR AGE

For Andrew Kimbrell

with warm regards,

PERSPECTIVES ON OUR AGE

JACQUES ELLUL

SPEAKS ON HIS LIFE AND WORK

EDITED BY WILLEM H. VANDERBURG

REVISED EDITION

ANANSI

English translation copyright © 1981 Canadian Broadcasting Corporation
Preface to the first edition © 1981 Willem H. Vanderburg
Preface to the second edition and appendices © 2004 Willem H. Vanderburg

All rights reserved. No part of this publication may be
reproduced or transmitted in any form or by any means,
electronic or mechanical, including photocopying, recording,
or any information storage and retrieval system, without
permission in writing from the publisher.

First published in 1981 by CBC Enterprises.
Published in 1997 by House of Anansi Press Ltd.

This edition published in 2004 by
House of Anansi Press Inc.
110 Spadina Avenue, Suite 801
Toronto, ON, M5V 2K4
Tel. 416-363-4343
Fax 416-363-1017
www.anansi.ca

Distributed in Canada by
Publishers Group Canada
250A Carlton Street
Toronto, ON, M5A 2L1
Tel. 416-934-9900
Toll free order numbers:
Tel. 800-663-5714
Fax 800-565-3770

Distributed in the United States by
Independent Publishers Group
814 North Franklin Street
Chicago, IL 60610
Tel. 800-888-4741
Fax 312-337-5985

House of Anansi Press is committed to protecting our natural environment.
As part of our efforts, this book is printed on Enviro paper: it contains 100%
post-consumer recycled fibres, is acid-free, and is processed chlorine-free.

08 07 06 05 04   1 2 3 4 5

CBC logo used with permission

NATIONAL LIBRARY OF CANADA CATALOGUING IN PUBLICATION DATA

Ellul, Jacques
Perspectives on our age : Jacques Ellul speaks on his life and work /
edited by Willem H. Vanderburg. — Rev. and expanded ed.

Translated from the French.
ISBN 0-88784-697-1

1. Ellul, Jacques. 2. Intellectuals — France — Biography.
3. Technology and civilization.   4. Church and the world.
I. Vanderburg, Willem H.   II. Title.

CT1018.E45A313 2003   944.081'092   C2003-903129-2

Cover design: Bill Douglas at The Bang
Typesetting: Brian Panhuyzen

**Canada Council Conseil des Arts
for the Arts du Canada**

ONTARIO ARTS COUNCIL
CONSEIL DES ARTS DE L'ONTARIO

*We acknowledge for their financial support of our publishing program
the Canada Council for the Arts, the Ontario Arts Council, and the Government of Canada
through the Book Publishing Industry Development Program (BPIDP).*

Printed and bound in Canada

Contents

Preface to the First Edition

WHILE DOING GRADUATE WORK in the applied sciences, I became increasingly struck by a strange contradiction between what we perceived the value of our research to be and the kind of world that emerged largely as a result of the application of this type of research. Countless highly specialized research efforts of the kind we were engaged in clearly led to advances in the means society uses for its existence. They helped to make things work better and more efficiently, and that could only be seen as good. Yet we were daily bombarded with information about the mixed blessings of science and technology. Obviously something unexpected happened as the results of countless highly specialized research efforts were woven together into new or improved means for our existence to become incorporated into the fabric of our civilization. But the training of researchers like myself did not prepare us to understand these processes, let alone anticipate or adjust for them in our work. The standard explanations to the effect that we only produced neutral means and that the problems arose because society applied them badly, did not at all satisfy me.

So I tried to find real answers by exploring the literature dealing with the way science and technology have affected past and present civilizations. I quickly discovered that much of this work was seriously lacking in a variety of ways. This can best be illustrated by using a simple analogy. If we wish to research the properties of water, we know better than to study only the properties of its basic components. The properties of water cannot be deduced from those of oxygen and hydrogen, because something new emerges when they combine. Under normal conditions, hydrogen and oxygen are gases, but water is a liquid with fundamentally different properties. We acknowledge this by saying that the whole is more than the sum of the parts. When it comes to the interactions between science and technology and the way they permeate society, producing new entities and structures, this is all too often forgotten.

It is frequently acknowledged that science and technology continually shape and reshape almost all aspects of contemporary society, but many disciplines pay very little attention to the roles they play. The high degree of specialization of these disciplines has not allowed them to see the massive new structures that have emerged. My search led me to the work of Jacques Ellul, who by means of his concept of technique has attempted to deal with some of these problems.

Throughout history human societies have created concepts to get at the meaning of the "wholes" in their experience. When in prehistory human existence was largely embedded in nature, people did not consider that this environment was simply a collection of constituents and phenomena that had no interrelationships apart from those that could be observed on the level of immediate experience. In other words, no group of people ever held that there were just rocks, trees, lakes, mountains, clouds, birds, and so on. Via science and religion, humanity has always asserted that this world has a nature and a structure.

While today we live in a largely artificial world, the situation is no different. We do not live on the level of immediate experience, and concepts such as the state, the economy, science, technology and industry have been created in order to make sense of our world. But many new phenomena have sprung up in the second half of this century, hence the question arises as to the adequacy of our stock of concepts for understanding reality. If these new phenomena indicate the emergence of new structures, new concepts may well be required to make sense of our world.

The creation of new concepts is a hazardous venture. If we go too far, we alienate ourselves by creating an ideology, while if we do not go far enough we barely go beyond immediate experience and fall short of understanding the forces and deep structures of our contemporary civilization. In either case, we will not be able to effectively cope with the challenges we face. We must therefore constantly put our theories and concepts to the test.

In creating the concept of technique, Jacques Ellul has made an important contribution toward understanding our age. In my opinion this concept may well become as central for understanding our times as the concept of capital became for the nineteenth century. Our world has emerged from what Ellul calls a technical intention, which is the preoccupation of our civilization with the one best way of doing things. It involves studying every human activity and utilizing the results to build some kind of model. By determining under which conditions the model functions optimally, one can proceed to restructure that activity to make it as efficient as possible.

The means to do so, in almost every area of modern society, Ellul has called techniques. These techniques are increasingly interdependent and have begun to constitute first a phenomenon, and later a system. Technique is clearly much broader than technology, which is only one of its branches. As it permeates contemporary societies, they are fundamentally changed, leading to an entirely new civilization. In his work, Ellul has examined many aspects of this civilization, showing us that our all-out attempts to render the means of our existence more efficient have produced something quite unexpected, of which other thinkers have seen only a part.

Just as nature presented so-called primitive societies with a variety of challenges, so does our new milieu. One of them, as Ellul points out, is the threat technique poses to human freedom. But we do not easily accept such a fundamental critique of our way of life because our existence is so inextricably bound up with it. To penetrate into the deep structures of our civilization is to expose some of the roots of our existence, and we instinctively defend ourselves to prevent this from happening. Yet it cannot be avoided if we really wish to understand our age. Questions of ultimate beliefs, political convictions, religion and faith will need to be addressed. Ellul does so by telling us something about his own life with its questions and difficulties, and his faith. In it we may be able to recognize some of our own questions and struggles as fellow human beings of this century.

There has been a tendency for translators of Ellul's sociological works to translate the French word *technique* as technique and to render it as technology in his theological works. I believe the former translation is preferable for two reasons. It keeps reminding the reader that what Ellul means by technique is quite different from what he means by technology, which is only one of its branches. Secondly, when we speak of specific techniques, the word technology is often inappropriate. We know what organizational techniques or techniques of public relations are, but we do not know what organizational technologies or technologies of public relations are. The translator of this volume, however, has preferred to follow the trend established by the translators of Ellul's theological works, and the reader should keep this in mind.

In closing, a word needs to be said about how this book came into being. In the spring of 1979 I was approached by Morris Wolfe to help prepare a series of radio broadcasts on the life and work of Jacques Ellul for the *Ideas* series of the Canadian Broadcasting Corporation. My main responsibilities were to prepare a plan for the programs, to interview Ellul and to be of some assistance in converting this material into a script. The first task was very difficult, not just because Ellul's writings are numerous and complex, but also because I wanted to give the audience a feel of the person as I had come to know him during four and a half years of postdoctoral work, when our two families shared a great many things in our lives. From the plans, I drafted a list of questions which became the basis for the first four programs in the series.

It was broadcast in the fall of 1979 and rebroadcast in 1980. The response was very encouraging, and the decision was made to publish the original interviews. Ellul helped convert the transcripts of the four talks into a manuscript, elaborating a few points at my request to yield the text, the translation of which constitutes this volume.

I wish to thank Jacques and Yvette Ellul for making this book possible. I also wish to thank Morris Wolfe for his support and helpful suggestions in planning the interviews and the large share he had in producing the *Ideas* program. My thanks also go to our producer,

Bernie Lucht, who was most patient and helpful in our first major radio venture, and to Carolyn Dodds of the CBC for her encouragement during the editing of the translation.

Willem H. Vanderburg
Toronto
January 1981

Preface to the Second Edition

MORE THAN TWENTY YEARS have passed since the first publication of this introduction to the life and work of Jacques Ellul. Subsequent developments in the world and much feedback on this introduction have prompted me to pass on the following additional remarks, in the hope that readers will find them helpful.

With the twentieth century now behind us, I believe that what the concept of capital did for helping us understand the nineteenth century, the concept of technique will do for our understanding of the twentieth century. The events of the past few decades have amply confirmed what Ellul foresaw in his first work dealing with technique. In French this work was entitled *La Technique*, with the subtitle *L'enjeu du siècle* ("The Wager of the Century"), and it was later translated into English as *The Technological Society*. Ellul's interpretation and expectations have been verified at a speed I certainly did not expect. Even the most recent world events can be explained by his concepts of technique, technique as life-milieu, and technique as system. The sad part of this is that Ellul would have preferred to be proved wrong, as a result of a decisive human intervention in the course of events—an intervention based on values and aspirations other than those resulting from the influence of technique. Instead, as technical developments continue in the future, a change of course will be increasingly more difficult, while the negative consequences of technique continue to become more severe and widespread.

In light of this situation, I am glad I decided against entitling this book according to the way Ellul summed up his life's work: "Think globally and act locally." That expression became a catchphrase for a number of causes and organizations, which emptied it completely of what Ellul meant by it. He was first and foremost an activist (in the best sense of this term) concerned about the many forces that undermined and threatened life in his time. His many writings, all prepared during the first two hours of his working day, were aimed at sharing his understanding of what was happening and what the implications were for human life and society, in order to intervene and hopefully contribute to steering things in a direction that was less harmful.

It should be noted that this task cannot be carried out within the boundaries of any particular discipline. Since there is no science of the sciences capable of *scientifically* integrating their findings, a different approach must be taken which incorporates, but is not limited to, science. Such an interpretation cannot be tested experimentally, nor can it be confirmed by the quantitative methods found in the social sciences and humanities. What you can do, according to Ellul, is to engage yourself in the world according to what the very best interpretation of what is happening reveals about what must be done to sustain life. If subsequent events bear out what you would have expected given these insights, the latter are confirmed in a small way, and you continue. If, on the other hand, subsequent events appear to contradict your insights, it is necessary to rethink the situation.

In all of this, it is essential to remember that we are people of our time, place, and culture: technique is not just "out there"—it is within all of us. I still remember clearly when, after reading a chapter and a half of *The Technological Society*, I was stunned by what I felt was the best implicit description I had ever read of how my mind worked as a doctoral student in engineering. The problems were "inside" as much as they were "outside." This was rather disconcerting, given that I was regarded as one of the radicals in the faculty because I insisted that the methods and approaches that had brought us to an environmental crisis and that were rushing us towards the limits of growth would have to be fundamentally altered. I was also astounded by how a sociologist and historian from another country and culture had so well grasped how my mind worked and what the consequences of all this would be. To think globally could apparently expose much of what we take to be obvious and given, as reflections of the influence of our time, place, and culture.

Ellul's many books and articles transcend one or more disciplines to form the "puzzle pieces" that fit together to create a narrative of the human journey with technique. This narrative integrates the scientific, technological, economic, social, political, legal, moral, religious, and aesthetic aspects of contemporary life, similar to the way they are experienced and lived. Much as rationality was regarded by Max Weber as a phenomenon larger than technology, so also the phenomenon of technique includes technology but is far from limited to it. The French language has no precise equivalent to the English word *technology*. Technology is taken to be only a particular expression of something much deeper and more fundamental. Just as capital was the "organizing principle" of the so-called capitalist societies of the nineteenth century, so technique is the organizing principle operating as a "system" in the so-called post-industrial and information societies of the latter half of the twentieth century. Technique has created an entirely new life-milieu, which has interposed itself between individuals and society, and between society and nature. As our primary life-milieu, its influence on human consciousness and cultures is possibly as great as that of nature in prehistory, and that of society in history until the twentieth century. Technique has permeated our language, values, morality, and aesthetic expressions, as well as creating new myths and a secular sacred. In this sense, technique may be regarded as constituting the "cultural DNA" of our time. Another way to view this situation is to recognize that as people create and develop technique, technique simultaneously influences people. If, from a historical and sociological point of view, the latter influence is more decisive than the former, a condition of reification is superimposed on that of alienation, robbing us of the margin of freedom essential for us to intervene in the course of events. This situation gave rise to Ellul's much misunderstood assessment that technique at this point in time has taken on a measure of autonomy with respect to human life and society. For these reasons, I deemed it inappropriate to translate the French word *technique* as the English *technology*, and I have made the appropriate changes in this second edition.

The life and work of Ellul are important for another reason. As a young intellectual, his life was turned upside down by an intervention of God. After his death, it became apparent that he must have destroyed the manuscript I know he wrote about this intervention. Ellul was horrified by what was done with the work of Karl Marx in the twentieth century: a "total" explanation of human history was turned into the basis for totalitarian regimes; and I suspect that Ellul feared the possibility of an attempt by the Christian right to abuse his work. As Ellul explains in this book, neither organized Protestantism nor Catholicism was of much help in coming to his understanding of what Judaism and Christianity could possibly mean in the presence of technique. I am keenly aware of the fact that many readers have difficulty grasping the interplay between his "sociological" and his "theological" works.

For this second edition, I have added two appendices designed to help readers avoid some of the widespread misunderstandings about the concept of technique and the relationship between Ellul's sociological and theological writings. In the first essay, I explain how technique may be understood as a way of interpreting and living in the world that both includes and transcends the culture-based approach by which human groups and societies lived previously. The second essay deals with how Ellul's work fits together. Based on twenty-five years of teaching this material to students in engineering, sociology, and environmental studies, I believe that these two essays can steer readers past the usual prejudgments and translation problems that could stand in their way. Ellul is neither a pessimist nor a "technology basher." Would you accuse the thermostat on the wall of your living room of bashing your heating system by constantly criticizing it in terms of the set-point that reflects your desired level of comfort? Without such bashing, regulating the heating system would be impossible. Similarly, intervention in the development of technique would be impossible without a constant assessment of technique in terms of human values and aspirations, provided these are not the result of the influence of technique itself. Labelling any critic as a pessimist or technology basher appears to be the result of a modern taboo on challenging science and technology, to which contemporary cultures have correctly attributed a very high value given the central role they play in human life and society. If nothing more valuable can be imagined, we are in the presence of a secular sacred, with the ensuing difficulty of regulating and dominating the creations to which an ultimate value has been assigned. In contemporary societies, anyone who seriously challenges technique risks being treated much like heretics were in the past. Thinking about this problem of secular idolatry can help us understand the relationship between the sociological and theological portions of Ellul's work.

Finally, many readers of the first edition have been baffled upon finding two versions of it in libraries and bookstores. One version has blank pages where the preface would otherwise be, reflecting the decision of the U.S. publisher to market it as a book written by Ellul. Readers may be assured that apart from the preface, both versions are identical. I hope this will clear up the mystery many librarians and readers have faced.

I trust that this second edition will entice and prepare my readers for digging into Ellul's writings.

Willem H. Vanderburg
Toronto
February 2004

PERSPECTIVES
ON OUR AGE

1
The Questions of My Life

I BELIEVE THAT AT the very start, one of the most important, most decisive elements in my life was that I grew up in a rather poor family. I experienced true poverty in every way, and I know very well the life of a family in a wretched milieu, with all the educational problems that this involves and the difficulties of having to work while still very young. I had to make my living from the age of fifteen, and I pursued all my studies while earning my own and sometimes my family's livelihood.

I was born in Bordeaux on 6 January 1912, but my family was not native to that region. My mother, who was also born in Bordeaux, was French. But my father was a foreigner—a complete foreigner, since my grandmother was Serbian, of the high Serbian aristocracy, and my grandfather was Italian. This factor, my father's background, is also very significant. One cannot imagine what it meant to be a Serbian aristocrat who was accustomed to very great wealth and a very easy life throughout his childhood, and then had to spend the rest of his life in extreme poverty after coming to France.

I was an only child and I lived with two parents who loved me very much, but in completely different ways. My father was very distant because he preserved a certain outlook from aristocratic life; my mother was very close to me, though extremely reserved.

In those days, I lived in great freedom, on condition that I fully respected the taboos, the orders, and the rules coming from my father. When he was absent, he didn't care at all about what I might be or what I might do. My mother gave me very great liberty, not out of indifference, but out of the conviction that freedom could be very fruitful for me. Since there were no distractions, the milieu I grew up in, outside of the lycée, was the port of Bordeaux—its wharves, its docks. I spent all my free days and all my vacations around sailors and longshoremen, with everything that this could involve—a totally astonishing milieu for a child; an environment that was very educational and, of course, rather dangerous, even though nothing ever happened to me.

Considering the part I eventually played in the French Reformed Church, people may wonder what sort of religious upbringing I had. I would say that in my childhood I really did not have any at all. I had none because my father, a highly intelligent and cultured man, was a complete Voltairian in both senses of the word. In other words, he was extremely critical about anything to do with religion and was convinced that it was nothing but myths, nothing but tall tales and fairy stories for children. Yet at the same time, he was utterly liberal— that is, he felt he had no right to coerce his son in either direction. Consequently, he did not want me to receive religious instruction, but he was not averse to my having some knowledge of the problems of Christianity. His religious background was Greek Orthodox, and, of course, the fact that there were no Greek Orthodox in Bordeaux made it well-nigh impossible for him to have any religious contact.

My mother, however, was a deeply religious Christian. She was Protestant. But out of loyalty to her husband, out of respect for his wishes, she never spoke to me about it; she never told me anything. Indeed, she never even went to church, though she was so very profoundly devout. It was only later on, when I began to ask her a few questions, that she revealed that she was a Christian. Hence, I originally had no religious upbringing. There was simply a Bible in the house, and it was one book among several. I did not live among great numbers of books, because we were poor, and I had no library at my disposal. This lack also deeply marked my childhood and my education, and it has helped me understand the situation of poor students.

There is truly a vast difference—even if one has lived in an intelligent milieu—between the person who had a family library at his or her disposal and the person who has never owned a book. All the books I had, I bought myself. The result, as I have sometimes told my friends ironically, was that, being a good pupil, I knew everything to be learned at the lycée, but I knew nothing else.

For example, in French literature, the curriculum stopped with Leconte de Lisle, and I knew everything about him. But when I heard some of my friends talking about Proust and Gide, I didn't even know the names; I didn't know who or what they were. No one had ever spoken of them in my presence, and that is a very important part of a youngster's education.

However, another family element was fundamental. My mother was an artist, a painter, and this contributed to the modest resources of the family, for she gave drawing and painting lessons. I believe she was a very good painter. If she had had the chance to exhibit her work, she would certainly have been successful. So, I lived in a certain artistic atmosphere—though exclusively one of visual art, which is also significant. I never heard music. I must have been twenty-three or twenty-four the first time I attended a concert. At first, I did not understand anything, because I was untrained. I was educated visually. I was accustomed to forms and colours, and I've always understood painting well enough.

Having done brilliantly at the lycée, I went on to higher studies. And here I can point out something that is characteristic of my mother. When I received my baccalaureate upon completing the lycée, well-meaning friends, who ran a large business, called on my mother and said: "Look, we know that your son has just gotten his baccalaureate, and we'd like to offer him a job. It would be a very good thing for him." My mother replied: "Never. He's got fine intellectual qualities, and he's got to go on to a higher education. Why, he's going to go all the way; we'll do everything we have to for that."

They did everything they could, and so did I. As for my choice of studies, I was not very good in mathematics, and literature did not lead anywhere. Law, however, was a subject that seemed to lead to a profession, and the study of it was relatively short. Those were frankly the only reasons I had for choosing it. So I entered the Faculty of Law and began my studies in the history of law and institutions. My doctoral thesis dealt with an ancient Roman institution, the *mancip-ium*. This was, essentially, the right of the paterfamilias to sell off his children. I defended my thesis in 1937, and I received my *agrégation* in 1943.

While attending the Faculty of Law, I encountered the thinking of Marx. This happened quite accidentally. In 1930, one of our professors of political economy was giving some economics courses on Marx. He aroused my interest, and I asked for *Das Kapital* at the library. I plunged into Marx, and all at once I felt as if I had discovered something totally unexpected and totally stupefying, precisely because it related directly to my practical experience. I also believe, of course, that it explained many subsequent events.

My father was a victim of the crash of 1929; he had lost his job. Hence, my family lived on whatever my mother earned as a drawing instructor and whatever I earned at any job I had then. I learned what unemployment is with no assistance, with no hope whatsoever, with no help from anywhere. I learned what it is to be sick with no government medical care and no money to pay the doctor or the druggist.

I remember my father spending his days looking for work. Given his abilities, I felt that it was an absolutely stupefying, incredible injustice that a man like him was unemployed; that he had to go from company to company, and factory to factory, looking for any job at all and getting turned down everywhere. And it was an injustice that I did not understand.

Then, in 1930, I discovered Marx. I read *Das Kapital* and I felt I understood everything. I felt that at last I knew why my father was out of work, at last I knew why we were destitute. For a boy of seventeen, perhaps eighteen, it was an astonishing revelation about the society he lived in. It also illuminated the working-class condition I had plunged into and those dealings in the port of Bordeaux, which I have already mentioned. Thus, for me, Marx was an astonishing discovery of the reality of this world, which, at that time, few people condemned as the "capitalist" world. I plunged into Marx's thinking with an incredible joy: I had finally found *the* explanation. As I became more and more familiar with Marxist thought, I discovered that his was not only an economic system, not only the profound exposure of the mechanics of capitalism. It was a total vision of the human race, society, and history. And since I did not follow any creed, religion, or philosophy—for I am very unphilosophical—I was bound to find something extremely satisfying in Marx.

We must not forget that at that time, in 1930, major things were happening in politics. Fascism was developing powerfully in Italy, and Nazism was beginning in Germany. Of course, being at the Faculty of Law, I was aware of these issues, but I knew them only from the outside, since France was not yet fully immersed in the conflict. However, despite everything, I discovered in Marx the possibility of understanding what was going on. I felt I had a deeper insight into the things I was being taught at the Faculty of Law. I was learning political theories, political science and constitutional law. But it all struck me as a bit shallow next to what I was reading in Marx.

Hence, I naturally made contact with people calling themselves Marxists. They were socialists, the socialists of the SFIO, *Section Française de l'Internationale Ouvrière* [French Section of the Proletarian International]. I must say that I was promptly and deeply disappointed, because I felt I was meeting people whose main concern was to "make it" politically. Apart from that, they had no real interest in transforming society.

I also met Communists. At that time, there were simply no Communist students in France. I tried to contact workers who I knew were Communists, and again I was very disappointed with their leaders, but for quite a different reason: whenever I started talking about Marx, they looked at me as if I were talking about something rather boring, and they launched into propaganda. What interested me, however, was Marx's thinking. I repeatedly tried to discuss it; but I was told: "Marx is not on our level, the important thing is the Party line." Well, that didn't suit me at all. Consequently, I never joined the Socialist Party or the Communist Party; I remained on the periphery. This was not yet the era when a number of brilliant French intellectuals entered the Communist Party. At that time, few intellectuals belonged to it. I stayed on the fringes, and I stayed there for a long time, until the period of the so-called "Moscow Trials," which began in '34/'35, and continued until '36/'37.

The Moscow Trials spelled my complete break with the Communists and the Communist Party. I feel that one did not have to be especially intelligent, especially enlightened, or amazingly lucid (as is now always said) to understand what was going on in the Soviet world. I

5

could not accept the charge that men whom I had read—for instance, Nikolai Bukharin, whom I profoundly admired—were traitors, that they wanted to destroy Communism and reinstate capitalism, etc. In other words, because I took Marx seriously, and because a Communist-type revolution had come about in the USSR, I could not believe that the people who had brought about this revolution were traitors to be rejected and condemned. Consequently, I told myself, one side must be lying, one side must be mistaken. Since I could not believe that the comrades of Lenin could have deceived and betrayed one another, then it had to be the Soviet government of Stalin.

On this basis, it was very easy for me to make up my mind. I was, in effect, led to reject Communism openly. I realized it was a totalitarian system. This shows, I believe, that for me Marxist thinking was not only on an intellectual level. That is to say, Marx provided an intellectual formulation of what, for me, had to come from experience, from life, from concrete reality.

I should perhaps explain the concept of dialectics. Dialectics, as a way of thinking and understanding reality, has become quite common, quite current in the Western world. This is due to the influence of Marxian thought and the rediscovery of the importance of Hegelian thought. I would say very simply that, at bottom, dialectics is a procedure that does not exclude contraries, but includes them. We can't describe this too simplistically by saying that the positive and the negative combine; or that the thesis and the antithesis fuse into a synthesis; dialectics is something infinitely more supple and more profound.

An example that may be easy to understand is as follows. We know very well now that there is no clear and evident opposition between life and death. Ultimately, every living organism has a certain number of forces working to preserve and renew it and a certain number working to destroy it. Hence, there are successive equilibriums between the forces of life and the forces of death. And the person or the organism evolves accordingly. Likewise, we can say that in every historical situation there is an aspect that might be called the positive one, and a contrary, contradictory aspect. But there can be no pure and simple elimination of the positive aspect by the contradiction, or of the contradiction by the positive aspect. That is, no exclusion occurs as it does in a logical process in which one says that white is the opposite of black, and that nothing can be both white and black at once. In a logical thought, the two are mutually exclusive.

In dialectical thinking, the contradiction is viewed as an historical development. Thus, the outcome is neither a confusion (white and black leading to grey, for instance) nor a synthesis in the ordinary meaning of the term. What happens is that a new historical situation emerges, integrating the two preceding factors with one another, so that they are no longer contradictory. Both have vanished, giving birth to a radically new situation.

This process allows us to understand an entire historical evolution, for example with Hegel or with Marx. But what strikes me as important is that Hegel and Marx did not invent dialectics. Yes, dialectics did exist among the Greeks. But for them, it was not a process of resolved contradiction, but something altogether different.

There is, however, another type of dialectical thinking, which inspired Hegel. This is Biblical thinking, both in the Old Testament and in Saint Paul. Here, we constantly see two contradictory, apparently irreconcilable things affirmed, and we are told that they always meet to wind up in a new situation. One example is Saint Paul's assertion: "You are saved by grace; *therefore* work for your salvation by your works." This sounds perfectly contradictory. Either one is saved already and saved by the grace of God; in that case, one need not bother working. Or one is called upon to work toward salvation with works, which means that one is not already saved and that one is not saved by the grace of God. Now Saint Paul says both things in the same sentence. This is dialectical thinking: once you are saved, you are integrated into history, into a process leading to your salvation, which is given to you in advance, but which you have to implement, which you have to achieve, which you have to somehow take in hand and utilize. But this cannot be done on an intellectual and schematic level. It will be done in the course of your life. That is why we are dealing with something contradictory, yet it is not contradictory when we live it. I can do the works necessary for salvation because I am saved. If I were not saved, if I were damned (assuming a notion that is not Biblical), I could not possibly do the works for my salvation. Hence, in the process of life, this is perfectly resolved.

In the same way, the Bible shows us, for instance in the Exodus, that God's people are set free only to be placed under God's control.

Are these two themes contradictory? Does God liberate in order to capture and reduce to slavery again? In other words, did the Jews leave one bondage only to enter another? Absolutely not. The Bible says that at this point, God, having liberated His people, controls them and guides them, but with the initiative and independence of this people. The people must constantly take up the conditions of their liberation again; and that is what the Israelites do. If we have been set free by God, then this means for our future. Hence, we must accept control and management from God at the same time as we accept this access to freedom.

Obviously, this is hard to grasp intellectually; and at the same time, it is something that can be lived concretely. Intellectually, it is the great problem posed by Karl Barth. Barth said that, on the one hand, there is the freedom of God and, on the other hand, the freedom that God gives to human beings. The goal is to live the human freedom within the freedom of God.

Thus, logically, the two cannot be reconciled. But dialectically, one can live them.

What turned me against the Communist Party was the difference between what I understood and what I then saw among the Communists. I believe that for me there had to be a coherence, a continuity between Marx's thought and one's life in terms of that thought. This was something I had not found in the Communist Party. The experience of the events in the Soviet Union alienated me completely from Communism, and my rejection of the Party, my total break, was confirmed when I saw what it did in the Spanish Civil War. One could say that the Communist Party was Franco's best support. Franco won the war because the Communists destroyed the resistance of the anarchists; their hatred of anarchism surpassed their hatred of Franco. And the same thing happened in the French Resistance during World War II. Many Frenchmen have said that the Communist Party was the chief party in the Resistance. But I saw the Communist Party wipe out Resistance centres for not being Communists. In our region, in March of 1944 I saw a Communist underground group destroy and kill all the members of a Gaullist group simply for being Gaullists.

Because of such experiences, I felt that the Communists no longer had the right to be heard, received, or believed. They really had nothing to do with Marx. And I feel that my experiences could have been those of anyone involved in the political conflict. As for the intellectuals who were Communists until 1968 and who now apologize and try to understand why they were Communists, I would simply say that they did not want to see what was happening.

I thus came back to Marx. I broke totally with the Communists but I drew close to the thinking of Marx, who unquestionably instilled a revolutionary tendency in me. I understood that the revolution would not be achieved by the Communists, and I was sure that the Nazis would not do it either. But I realized that the world I lived in was a world that could not go on as it was indefinitely. The issue of revolution was central in my youth, and it has remained central to me throughout my life. It was Marx who convinced me that people in the various historical situations they find themselves, have a revolutionary function in regard to their society. But one must understand exactly which revolution it is; and in each historical period one must change, one must rediscover. This was an element that Marx planted in my life and that has never changed.

Another element, certainly, was the importance of reality. (I am not speaking of materialism.) Marx assigns major importance to the concrete material reality that surrounds us. Both the intellectual and the spiritual minds tend to forget this reality, to disguise it, as though it could ultimately be masked. But because of Marx's influence, whenever I speak, I instantly ask myself in terms of what economic situation I am speaking, what my interests are. This too was something that deeply marked me in regard to myself and, obviously, in regard to everything around me.

A third element of Marx's influence, of course, was my decision to side with the poor. But here, one must be careful. Marx was extremely precise about the poor. For him, the proletarian was not just one who is poor in money. This is shown, for example, by his lack of interest in peasants, who are poor in money. For Marx, the proletarian is the person who is alienated by all the modern conditions of life. The proletarian, the true pauper, is subjugated to the imperative of the machine and lives in the city, uprooted, in an unacceptable urban condition. Proletarians cannot have families because their economic conditions prevent them from living a family life. Contrary to a statement in the *Communist Manifesto* of 1848, and contrary to what is normally believed, Marx was not opposed to the family. He himself started a family and was a good father who married off his daughters and so on. Hence, Marx was not hostile to the family per se. He was hostile to the fact that the bourgeoisie has turned the family into a privilege. In other words, the unacceptable element of capitalism is not the existence of families, but that certain people can have a normal, happy family, while others, a majority, cannot. Marx's ideal, however, is that a person should have a family and that the parents should be a happy, balanced couple with happy, balanced children.

Consequently, the poor person is the person who cannot have this family. For Marx, there is a complete analysis of the psychological, sociological and economic situation of human beings, and the poor person is the person deprived in all these areas. Hence, when I say that Marx oriented me toward always siding with the poor, I am not necessarily siding with those who have no money. I am siding with people who are alienated on all levels, including culturally and sociologically—and this is variable. I will not claim that qualified French workers in the highest category are poor, even though they are subject to the capitalist system. They have considerable advantages, and not just material ones. On the other hand, I would say that very often old people, even those with sufficient resources, are poor, because in a society like ours they are utterly excluded. That is why I have sided with the excluded, sided with the unfit, sided with those on the fringes. That is why I keep discovering those who are the new poor in a society like ours.

Since we are talking about the influences that Marx had in the system of my life and thought, I would like to emphasize that in the religious area or in regard to the Church, Marx had no influence at all, for the good reason that I was not particularly touched by his arguments about religion and God; these topics did not really interest me. On the other hand, his great attacks against Christianity and the Church did not affect me because I was totally indifferent to the Church. It existed vaguely on the level of Catholic rituals, but that was just about all. I was given to a certain skepticism and—obviously emulating my father—to a highly ironic attitude about the fasts and ceremonies of the Catholic Church.

I have said there was a Bible in our home. I did read passages and texts in it, and many things interested me—even, I would say, seduced me. But I was almost more seduced by the Old than by the New

Testament, by the stories, the prophets. Even when reading Marx, I was brought back to the social and political proclamations of the prophets, which struck me as interesting.

I was inspired to ask questions, but my father did not reply, and my mother gave me simple, elementary answers. Furthermore, when I asked slightly more precise questions, I received no answers. Our Bible had some words in italics. Today, I know why. The translator was very conscientious, and he put all the French words that did not appear in the Greek or the Hebrew text in italics to show that he had added them. But when I asked my mother, she did not know why they were there, nor did my father. Mother said, "Well, go and ask the pastor." I went to the pastor and asked him, but he did not know either. Several times, I had questions—I could cite many of them, a child's questions—and each time, I was sent to the pastor, and the pastor did not know.

"Fine," I said to myself, "adults can't answer my questions. I'll have to get along on my own." This was very helpful for the later development of my mind; I ultimately did everything by myself. Every time I came up against an intellectual, philosophical or scientific difficulty, I no longer sought out authorities. I never looked for anyone to explain anything to me. I worked on the problem until I could explain it to myself. That was the basis of a whole intellectual attitude.

When I was faced very concretely with the question of death—which I had avoided as a child, even when I saw tragedies on the docks—I quickly realized that Marx did not have answers for everything. There were existential questions, the questions of life and death, and also—I was eighteen or twenty—the question of love. In regard to life itself, a certain number of problems were still open. It was here that the Bible gave me more, establishing itself in my life on a different level than Marx's explanations about society. In the Bible, I was led to discover an entire world that was very new to me because I was not accustomed to religious discourse, Christian discourse. A new world when I compared it with the realities of life and of my life and experience. I was converted—not by someone, nor can I say I converted myself. It is a very personal story, but I will say that it was a very brutal and very sudden conversion. I became a Christian and I was obliged to profess myself a Christian in 1932. From that moment on, I lived through the conflict and the contradiction between what became the centre of my life—this faith, this reference to the Bible, which I henceforth read from a different perspective—and what I knew of Marx and did not wish to abandon. For I did not see why I should have to give up the things that Marx said about society and explained about economy and injustice in the world. I saw no reason to reject them just because I was now a Christian.

This was not an era when such an attitude was easy. First of all, it was not common to deal with Marx back then. Unlike today, there was no group of Christian Marxists, who believe that it is very easy to be both a Marxist and a Christian. For me, trying to take both seriously was very difficult. I was forced to face the things that Marx said about religion, God and Christianity. I was obliged to accept this criticism, and I accepted it on a concrete level. Basically, Marx told me what a Christian should not be. He violently attacks the "pseudo-Christians" of the nineteenth century, but he absolutely does not reject the Biblical teachings. Hence, at that time, Marx played a very great part in my life. Conversely, the Bible did not make me reject the things that Marx said about economics and politics. I was thus placed in a contradiction because I did not create two separate domains. I realized that Christianity was a totality implying an ethic in all areas, and that Marx too claimed to be a totality. I was sometimes torn between the two extremes, and sometimes reconciled; but I absolutely refused to abandon either one. I lived my entire intellectual life in this manner. It was thus that I was progressively led to develop a mode of dialectical thinking which I constantly made my foundation. In my own life, I confronted the demands of Marx and the demands of the Bible and put them together. I did not create two domains. I did not say that there was the material on one side and the spiritual on the other. This opposition is absolutely wrong, even from a Christian viewpoint. After all, there is the incarnation of Jesus Christ, where God becomes flesh, or material. Likewise, I rejected something that Marx said: "There are an infrastructure, seriousness, solidity, economy, technology, and then there are imagination, philosophy, theology, which are not important." For me, the two elements were to be kept together, and it was necessary to progress stage by stage and with this permanent contradiction. That was ultimately the key to all my later thinking.

Marx changed several aspects of the way I read the Bible. I could not approach the Bible like an innocent Christian reading the text in its simplicity. Marx taught me—this is no longer new, but at that time, it was new—that a text comes from a certain milieu, from a certain economic situation. Consequently, I was compelled to read the Bible with a critical view deriving from that idea. I absolutely could not divorce the Biblical demand from the concrete economic or political reality. For me, the two necessarily went together.

Marx brought me into another predicament. I had converted, I had tried to get close to the Church, and I was not all that enthusiastic. Marx forced me to study the Church sociologically. Here (I am taking a chronological leap), I might say that one of my first important pieces of work was a sociological study of the Church I undertook when I took part in the World Council of Churches. In particular, I presented a long investigation of the sociological causes of ruptures in the Churches, which was not very well received. At any rate I had to make a critique of the Church very early on, because Marx gave me the intellectual instruments to do so and to view the established Church as a sociological entity that can be analyzed sociologically. The established Church is not the equivalent of the body of Christ.

In fact, however, I entered a Church because one cannot be Christian all alone. I made some efforts toward the Catholic Church in 1932–33. It did not greatly excite me. I felt, especially with friends of mine, that Protestantism was closer to the Bible. The group of Protestant students (the *Fede*) was very lively, very authentic—the cream of the student body. They managed to convince me. That was how I came to join the Reformed Church in France, a Church that was only faintly Calvinist at that time.

Bear in mind that we were going through the period of so-called liberalism, that is, a very rationalist and reasonable attitude toward religion and the Bible. I found myself among people who, for example, read the Bible, but did not believe in miracles. They saw Jesus as a good moral model, but felt that his resurrection was obviously a fable. Hence, I knew almost no real Calvinists in the Church. I discovered Calvin's ideas in a very small group. At that time, his thinking certainly beguiled me with its rigour, intransigence, and total use of the Scriptures. Thus, I read him enthusiastically and went along with his thinking for a certain time. But then I broke away from him.

I should stress that throughout the period of my study of Marxism and my commitment to Christianity and the Reformed Christian

Church, I pursued my studies at the Faculty of Law, I obtained my diploma (a *license*), and I began a doctoral program. I believe that the discovery of Christianity greatly influenced my understanding of law itself and my choice of a profession. For instance, in the state that I was in, that the society was in, I considered it impossible to become a judge. Both as a Christian and as an adherent of Marxist thought, although not a Marxist, I could not possibly become a faithful servant of capitalist society. It was out of the question. Nor could I enter any of the countless administrative or legal professions opening before me. It was a difficult predicament, and I finally chose teaching—first of all, because I had the impression that my intellectual life would allow me to teach facts about life to my students. Secondly, I had the impression that the function of teaching was one of the most detached, one of the most disengaged from the demands and imperatives of the world we live in.

It was through meetings with fellow students (and also two or three young Protestant theologians, one of whom, Jean Bosc, became my great friend) that we discovered the works of Karl Barth. Barth then became the second great element in my intellectual life, completely effacing Calvin. Obviously, once I began reading Karl Barth, I stopped being a Calvinist—in my understanding of the world and politics as well as, theologically, in my understanding of predestination, original sin, and the question of universal salvation. Obviously, I could no longer be a Calvinist once I understood the dialectical movement of Barth's thinking, which also brought me a clear and, I would say, free view of the Bible. Barth had an extraordinarily liberating effect, offering a method of comprehension far more than solutions. Calvin constantly offers answers, solutions, or a construction, while Barth launches you into an adventure. At the same time, I found myself dealing with a dialectical thinking that was fully consistent with what I had so vividly experienced, so vividly accepted and admired in Marx. Barth was a signpost showing how one could get beyond the stage of pure and simple contradiction between Christian faith and Karl Marx. For me, the implications of Barth's thinking are still not exhausted—far from it. I am always astonished and wax a bit ironical when someone declares that Barth is old-fashioned, that we are now in a post-Barthian theological situation. We are very, very far from drawing all the ultimate conclusions, especially on an ethical level, from Barth's thinking. For many years, part of my work has been devoted to seeing the significance of Barth's theology, which people claim is now obsolete.

After that period, that is, after 1933, when Hitler seized power and tragic times came for most of us, I got very deeply involved in politics. One of the most memorable times was February 1934, which brought the first great riot in France since World War I and the riots of 1917. There was the Fascist riot of 6 February 1934, where I was part of the crowd. On 10 February 1934, we gathered in Paris to fight the danger of a Fascist coup d'état, and 1935 brought the Italian attack on Ethiopia. This generated the first anti-Fascist movement, and I took part in it. There were the great right-wing strikes to force the government to support Italy. We had paltry means to fight them, and I fought a lot in 1935. In 1936, I participated in the Popular Front. Then came the Spanish Civil War, in which I took a modest part until late 1937.

For me, this was a period of very intense political involvement. During this time, we launched the personalist movement under the leadership of Emmanuel Mounier. There were about ten of us at first. We based our movement on a Christian foundation, Mounier was Catholic; but our political position was difficult. We were neither Stalinists nor Fascists; nor did we support liberal capitalism. We tried to get beyond this contradiction to surpass it. The movement was an extremely intense, extremely fruitful meeting place. The personalist movement was based on a philosophy that was altogether new, a philosophy that rejected individualism, which itself is deeply attached to the nineteenth-century bourgeoisie. However, we also rejected collectivism. We felt that a human being is a person, which means an economic as well as a spiritual being. We believed that a society must be structured purely toward developing this personhood and rejecting alienation. But on the other hand, one can be a person only if one belongs to a group, only if one belongs to a community. We followed a sociological viewpoint that is now criticized, but which struck me as very fruitful in opposing the community to the collectivity, the group of close relationships to distant relationships, basing everything on the importance and decisive character of the interhuman relationship. These, you see, were very modern motives and also, quite obviously, very dated motives of a certain philosophy and a certain political experience. And we nearly succeeded, I would say, we nearly succeeded. But we were quickly overtaken by the events. We belonged to no political party, we were chiefly anti-Fascist, and we might have given birth to a new and different politics. But then came 1938, and then 1939 and the war, in which everything disappeared.

During 1938 and 1939, I applied for the *agrégation* competition. I already had my doctorate, which made me eligible, and I had a job as lecturer at the Faculty of Law from 1937 to 1939. The years 1939 and 1940 were, of course, very hard for us, as for all Frenchmen. At the time of the armistice, when Marshal Pétain came to power, I was in the city of Clermont-Ferrand. I was an assistant lecturer at the University of Strasbourg, and the university had withdrawn to Clermont-Ferrand, where I joined it after the defeat.

A few days after the takeover by the Vichy government, I learned that I had been dismissed. There were two reasons. First of all, in the midst of the defeat, I had found myself among Alsatian students. They were utterly bewildered, anxiety-stricken, and fearful, they didn't know what would happen to them now that Alsace was to be incorporated into Germany. Since I was generally popular with the students, a group of them who were registered at the faculty had stopped me and asked: "What's to become of us?" As I often did, I gave a small political talk, telling them not to believe anything that Marshal Pétain said and, above all, absolutely not to return to Alsace, where they would obviously be drafted into the German army. I told them that the Germans would use any means to get them back to Alsace, which turned out to be true. The Germans subsequently made the parents of these students come to Clermont-Ferrand and ask them to come back to Alsace with them. I told the students that they simply mustn't give in to the German demands. I gave this speech to some fifty or sixty students, and one of them reported my statements to the police. (His fellow students told me who it was, and I later found out that the unfortunate boy was indeed conscripted into the German army and died on the Russian front.)

I was summoned by the police. The police commissioner, who was basically not very much in favour of the Vichy government, was very nice. He said: "You know, what you did was pretty irresponsible. I ought to arrest you; speeches like that are defeatist. But, okay, keep calm, I won't say anything." Nevertheless, the report reached the ministry, and they also discovered in my file that my father was foreign-born. So, according to the laws of Vichy, there was a double reason to dismiss me: for being the son of a foreigner and for having made hostile statements.

So I was dismissed. When the *agrégation* competition came up again, I was authorized to take the examination in 1943, but with a handwritten note from President Laval saying that if I were received, I would in any event not be admitted to the university. How encouraging to know this at the start of a competitive examination!

My wife, who is of English nationality but Dutch birth, had just had a baby. After my dismissal, we left and went back to Bordeaux, in the occupied zone, because we had no reason to go anywhere else. But upon arriving, I learned that my father had just been arrested by the Germans and that my wife was among a number of foreigners soon to be arrested as well. Not having much choice, being without a job or any resources, I vanished into the countryside with my wife. Students often ask me: "How did you join the Resistance?" I explain that it was not by choice. Where else could I have gone? I had just been dismissed, my father had been apprehended, and my wife might be arrested. I had no other option.

I was forced to join the Resistance. It was necessity, not virtue. We moved into the countryside and lived fifty kilometres from Bordeaux in a very isolated area. We were very well received by the farmers, who never asked us any questions. They were extraordinarily helpful, and I became a farmer and remained one throughout the war.

I raised sheep, with the help and advice of my neighbours. In 1943, when I finally brought in my first ton of potatoes, grown entirely on my own, and a bumper crop at that, I was as proud as when I had passed my *agrégation*! I supported my family completely with my farming. This included my mother, my father having died in 1942. I also began to participate in the *maquis*, the Resistance groups. I took part, according to my means, either in a discussion of tactics (which is not unimportant in such movements) or as a liaison agent, circulating rather easily among various groups. Throughout that period, I lived almost totally isolated from my former interests and activities. Yet even in this isolation, I was able to establish contact with one aspect of my former life.

During outings for the *maquis*, I discovered an abandoned church, and I learned that there was a Protestant community here, which had no leadership. I visited a number of these Protestant farmers, who, in 1943, resumed regular worship in this church, under my direction. This was very enriching from all points of view. I found myself among farmers, who, being generally people of few words, and never making any grand political declarations, revealed a spirit of resistance toward Germany and toward the right wing which I found surprising, even extraordinary, for farmers. No one had ever mentioned the Resistance to me before, but early in 1944, a group of farmers came by and said: "Listen, we know you're in the Resistance and you can count on us if you ever need marksmen, we're all yours. *Voilà.*" Such experiences are hard to forget in a relationship with very plain people.

In 1944, at the Liberation, I was part of the Movement of National Liberation, I even held certain positions in it, and had begun to believe the dream we had been dreaming during the last few years of the Resistance, often expressed by the saying that we were going to move from Resistance to Revolution. But when we said that—and I would like to point out that Camus first used it in 1943 in combat groups—we did not mean a Communist, Stalinist, Soviet revolution. We meant a fundamental revolution of society, and we made great plans for transforming the press, the media, and the economic structures. They all had elements of socialism, to be sure; but I would say it was more of a Proudhonian socialism, going back to grassroots by means of a federative and cooperative approach.

We believed these things. And when we surfaced during July 1944, we were several small groups trying to launch a revolutionary action. But then we were blocked again—on the one hand by General de Gaulle, who wanted to install a traditional republican government; and on the other hand by the former political parties, who instantly rose up again. The Communist Party, the Socialist Party, and the Radical Party reenlisted their former supporters and their old organizations. In the midst of all this, we were obviously very weak. Still, the best of us got politically involved—I went on until 1947—but we were profoundly disappointed. Some of my books, for instance *The Political Illusion*, derive in part from my experiences in the political milieu— from politicians' inability to really change the world they live in, the enormous influence of administrative bodies. The politician is powerless against government bureaucracy; society cannot be changed through political action.

As a result, I abandoned political life since I was not seeking a career—I was not trying to become a politician or a minister of state. I could have become one right after the Resistance; it was not very difficult. But that was not what I wanted. I learned that the revolution had failed once again. The Popular Front failed and I witnessed it; the Spanish Revolution failed and I witnessed it. The Resistance failed to turn into a revolution. After these experiences I did a critical analysis of revolutionary movements, and I had to ask why each revolution ultimately fails. This also explains some of my books, for instance my two works on revolution.

After that, what could I do? I resumed my profession, at my university. Naturally, I was welcomed back with open arms. At that time, it was an honour to have been dismissed by the Vichy government. As I entered the university, I thought of creating a parallel university, that is, next to the official one, with the best students. I wanted to organize small study groups to think critically about things and not just toe the traditional line. I felt that this would be a way of modifying the present and future structures of society. I believed, however, that a parallel university should not just be the same as the regular university but with different courses. We would have to live differently. We would go off to the mountains for two or three weeks, a month, two or three months. We would have an intensive course of studies and a hard life in the mountains, and we would do both practical work and critical reflection on society together with the students. I pursued this project as long as I was not too old to camp out with my students.

Then I launched into another enterprise. I said to myself that if there are any people capable of changing the society they live in, then it would be the Christians. I had my Christian reasons for transforming this society—why not work with other Christians? Why could I not get the Church to change and become the salt of the earth, a leaven, a force that would change society? So I decided to pursue a career in the French Reformed Church; and I attained the highest position in the Church: I was part of the National Council. This Council is a group of twenty people, ten pastors and ten laymen, who direct the Reformed Church. But I realized that the Church would have to be changed, in order to become a leaven, a force to change this society. So I began to try to change the Reformed Church. I worked at it for fifteen years. It was a difficult job, requiring tremendous patience—

and ultimately, I failed. We were able to set up a number of task forces, and we even got the Church authorities to accept certain new directions. But in the end, we bogged down in the traditionalism of Christians, in a sort of indifference toward change and in the institution itself. Once a movement becomes an institution, it is lost.

Finally, I felt that the study of theology would have to be changed. And in this, I succeeded; Protestant theological studies in France are different from anywhere else. But it is still too soon for me to see the results. I kept trying to find what would be possible for a Christian who analyzes society with the apparatus of Marx's thinking.

In my efforts, I looked for areas of action that would show where one could have an effect on this society and I found two. The first was the prevention of juvenile delinquency by working with so-called social misfits—the delinquents or predelinquents. I began this activity around the time of the *blousons noirs* [French Teddy boys—Translator's note], in 1958. It is still going on, but I stopped being directly involved in 1976. By then, I felt that we had done some grassroots work with many young people, and helped the government, the police, and the legal system to understand that when we speak of the socially maladjusted, we must first ask ourselves whether it is the individual or the society that is maladjusted. In other words, is it the young people who do not fit in or is it the society? We spent twenty years making people understand this, and I believe it is beginning to penetrate. People no longer regard violence, drugs, or hippyism as diseases of the young, but as diseases of society. We really worked hard on this challenge, which is a fundamental one, and it is more or less accepted now, even though we must never believe in victory in these areas! In fact, in France, we are witnessing a spectacular reversal in government policy in dealing with the problems of prevention. The government now rejects our movement and is cracking down on the young.

And finally, my last activity, my last involvement was my ecological commitment, a commitment to the environment. This was fully in keeping with all my research on the society dominated by technique and the present influence of industry on the transformation of the human environment. Hence, I was automatically in favour of working for ecology—the defence of the environment. Since 1968, we have formed several groups interested in the cause. For me, it's always the same problem. *Intellectual interest means concrete commitment,* practical and political involvement. So we launched ourselves into the problems of nuclear energy and the national planning of land use. The latter is controlled by an enormous bureaucracy. In the southwest, for instance, in Aquitaine, where I live, we are faced with an extraordinary lie. The authorities keep telling us that they are going to develop the maximum of tourism while protecting nature as much as possible. This is a lie, and we mounted a hard attack on the administration. This was very interesting for me, because in this ecological struggle, which was very concrete and not at all theoretical, I attacked the three things I despise the most: technique (let us say, technicists), bureaucracy, and capitalism. Aquitaine was a place where the three combined perfectly to fully destroy the Aquitaine environment and the Atlantic coast.

When I say that I "despise technique," I should perhaps explain. It is not technique per se, but the authoritarian power that the "technicists" seek to exercise, as well as the fact that technique determines our lives without our being able to intervene or, as yet, to control it. It is in these terms that I attacked technique in this local action. But only in the very important sense that technique was becoming a justification for unnecessary work. Bureaucrats and owners of capital were using the argument of technique to force acceptance of their projects. Another feature was the great number of technical *errors* and technical misapplications, all of which led nowhere.

I may give the impression that on the one hand I think in a rather general, rather all-encompassing theoretical manner; and that on the other hand all my efforts and involvements are on a small scale, whether in the World Council of Churches, in the teaching of theology, in the prevention of juvenile delinquency or in ecology.

I would say that this is quite true. I do not believe in global actions. I do not believe in actions on the level of the president of the Republic. In fact, I always apply a motto: "Think globally, act locally." And this corresponds to what I have tried to be. By thinking globally I can analyze all phenomena, but when it comes to acting, it can only be local and on a grassroots level if it is to be honest, realistic, and authentic. I also believe that this is a challenge to all the processes of action that are generally employed in our society.

To summarize my position in relation to Marx, the Christian Church, and the state, it is obvious from what I have said that I have great admiration for Marx, and I am quite influenced by his thinking. However, I remain extremely critical on two levels. First of all, in regard to Marx's assumptions, his thinking is not scientific, it is passionate; and it interests me for that reason. I certainly don't claim to be any more scientific than Marx, my thinking is passionate too. But what I would criticize in him is that while he was very clear-sighted about prejudices and presuppositions in everyone he attacked, he nevertheless failed to see his own. It seems to me that the first rigorous step one should take is the critique of one's own biases. In particular, Marx was the victim of two prejudices of his era. First of all, the prejudice of progress: he believed that every historical stage was an advance over the preceding stage. Secondly, the prejudice of work: he believed that work is what essentially characterizes the human race. These are the elements I criticize in Marx. I likewise criticize him for being convinced that he had answered certain questions, which he did not, in fact, answer. For example, in regard to the ultimate meaning of human life and of history, he keeps stating that he has found an answer, whereas, in reality, he never offers one.

Those are my two reservations about Marx. And if I have sometimes attacked him rather stoutly in my works, then the reasons were purely a result of the circumstances. Living in France, where practically all intellectuals are Marxists, I find it extremely difficult to be both faithful to Marx's thought and different from these intellectuals. Consequently, my positions are far more matters of differentiation from the movement of Marxist intellectuals than critical attitudes toward Marx himself.

As far as my involvement with churches is concerned, I was drawn to this work for two reasons. First of all, I felt that nothing less than the strength of the Holy Spirit could help us out of the crisis of our civilization. I believed that Christians would have sufficient motives for committing themselves to changing society, that is, a total revolution. However, I could reach Christians only through some organization, and I preferred the Reformed Church because it was weaker, because it was very unorganized, and because I felt I could reach people more directly there. But it was necessary to transform the hierarchy of the Reformed Church. Unfortunately, I was finally convinced that the sociological weight of the hierarchy is always the same. That is, even when this hierarchy affects only a small number of people, even when it is minimal, it is the inertia of the hierarchy that ultimately wins out over any desire for reform. I had hoped that I could have a highly flexible organization, allowing me the direct expression of personal relations. However, I had not sufficiently analyzed the Reformed Church, as modest a sociological structure as it may be. Hence, my critique of the machinery of the state and French politics after 1944 were not enlightening enough to prevent my involvement with this work in the Reformed Church.

As far as the state is concerned, I have participated in several national government commissions on such problems as the question of violence and the prevention of juvenile delinquency. On certain questions I am consulted as an expert, and I must say that I never refuse the government when it comes to me, even though I realize the results will not be very great. This was proven by the Commission on Violence; we produced excellent reports, but they had no effect.

2
Understanding Our Age

My political activity and my reflections based on Marx's thinking led me to establish two rather simple things. First, a good number of Marx's predictions about the evolution of capitalism did not come true. The transformation of the world was far more complex than he had envisioned. The capitalist world had powers of resistance that were not exhausted, despite Lenin's explanations. Second, a very large number of those bourgeois that Marx had talked about disappeared, especially the ineffective and useless portion of the bourgeoisie, the people limited to existing on a private income. Thus, there had been a certain transformation of capitalism too. I therefore wondered if the Marxian analysis of capital and capitalism in the nineteenth century was equally valid in the first third of the twentieth. It was certainly questionable. Next, it struck us, especially in the personalist movement, that there were certain extremely similar trends in both Soviet and capitalist society. Beyond the economic transformations and beyond the political and legal forms, one could find common elements—particularly the need to increase industry at any price and to develop technical objects.

Here, too, we were left with a question. We felt, perhaps because we hadn't read all of Marx, that he hadn't given technique the position it has in our era. The first person, no doubt, who stressed the importance of technique (in the proper sense, which I will try to define below) was my friend Bernard Charbonneau. In 1934, he began to regard technique as the decisive factor, the essential factor in the world we live in—a truly prophetic view. But Charbonneau, who teaches geography, did not create the stir that his ideas merited. He is completely unknown despite his highly remarkable books. It was he who first drew my attention to the phenomenon of technique. I gradually realized that a transformation had indeed taken place since the nineteenth century. Basically, Marx was speaking of a society dominated by the industrial world. In 1930 to 1940, this industrial world was still dominant. But now new trends had emerged.

It struck me that something similar and comparable in both the Soviet and the capitalist worlds was precisely the technical phenomenon. One could start with the extremely simple idea that a Soviet and an American factory were exactly the same thing, just as a Soviet and an American automobile are the same thing. In other words, on a totally elementary level there were common points, and this was a reason to compare the two kinds of organization. Little by little, as we analyzed the influence of technique and its importance in our society, we came to realize that technique was the most decisive factor in explaining our era. As an explanatory element, it could play the part that capital had played in Marx's interpretation during the nineteenth century.

I don't mean to say that technique has the same function as capital. Nor am I saying that the capitalist system is a thing of the past. I know that it still exists, but capital no longer plays the role it did when Marx was studying it in the nineteenth century. Power in particular and the reproductive capacity of value are no longer tied to capital; they are now inherent in technique.

However, perhaps we ought to be more precise. When I use the French word *technique*, normally translated into English as *technology*, I do not mean exactly the same thing as the French word *technologie*, which is also translated into English as *technology*. We have to be meticulous about this simple point of vocabulary. I know that the two are habitually confused. Etymologically, of course, *technologie* means a discourse on *technique*. That is the true meaning of *technologie*. Now when I speak of *technique*, I am speaking of the technical phenomenon, the reality of the technical. When I view an automobile, the engine of the automobile is in the category of *technique*, i.e., the technical. It is not what the French call *technologie*, even though English usage tends toward *technology* on this point. The study of the engine and the discourse on the engine is *technologie*. But the phenomenon itself must be viewed as part of *technique*. I know the difficulty of this semantic problem in English, for there is only one single word, *technology*, to designate both *la technique* (the concrete thing) and *la technologie* (the discourse, the teaching of the subject itself). But we must absolutely distinguish between the two. It is the same difference as between *society* and *sociology*, or between *earth* (*gē* in Greek) and *geology* (the science of the earth). However, there is a further difficulty. The English word *technology* essentially concerns the work of engineers, chiefly in the industrial milieu. But for me, *la technique* is a far wider concept, referring to efficient methods applicable in all areas (monetary, economic, athletic, etc.). I would prefer that English retain the word *technique*. Thus, in this sense, it is *technique*. In this reality, in this substance—one might say in our Western society—it is *technique* that struck us as the determining element, and also as the determining element in the creation of, say, value.

We know that for Marx, work is what creates value. We are bound to see that in a society which has become extremely technical, the determining factor is both scientific research and the application of science in the form of technique. These statements are not peculiar to the capitalist structure. This is what creates value now; even some (though not all) Soviet and Communist economists acknowledge it. In other words, we have to reread the world in which we now live. Not in terms of the capitalist structure, but in terms of technique.

The further I advanced, the more I asked myself which phenomenon would have struck Marx as the one most determining our society if he had worked in our twentieth-century milieu instead of the nineteenth-century milieu? Which phenomenon would have struck him as the one most crucial to structuring both the human condition and the political organization? I grew more and more convinced that technique is the element that would have caught his attention. Hence, it was in terms of Marxist thought and with relative faithfulness to Marx that I began to study the phenomenon of technique more and more closely.

Of course, others had more or less discerned the role of technique. I am thinking of Max Weber in particular, and then Lewis Mumford a bit later. But I feel that one cannot fully compare my research to theirs. In Max Weber, we most certainly have a very closely related method, but I cannot say that Weber influenced me. When I commenced these investigations, I was totally unacquainted with Weber's sociology, and I didn't get to know it until 1944. We certainly have a similar approach to issues and a similar sociological method, but there is a major difference between us.

However much of a genius and prophet Weber may have been, the society he analyzed was the society of 1900, or at best, the society of the nineteen-tens and -twenties. He died in 1920. Hence, he did not know the technical phenomenon in its full development. Scholars now generally agree that the watershed between the older society and the typical society dominated by technique came around 1945. In other words, Weber had a particular view of how general the technical phenomenon was. He thought about the bureaucratization of society in terms of technique, but he could not really study the phenomenon himself.

However, many other sociologists have studied certain aspects of our society dominated by technique. I am thinking of Raymond Aron in France and Galbraith and McLuhan in the United States and Canada. Aron has essentially studied industry; Galbraith, the technical, bureaucratic, industrial state and a particular power structure—the technobureaucracy, as it were; and McLuhan has studied the problem of mass media. But all of them, in my opinion, have done only fragmentary research. One cannot investigate the whole modern social phenomenon on the basis of the technostructure any more than on the basis of television. In other words, there is a general, over-all view encompassing research on industry, the modern state, and television. This all-inclusive view, this framework, is that of technique. Raymond Aron is very critical about some of my research, finding it much too general and systematized. But it is systematized only in that I try to offer a theoretical explanation for a phenomenon that strikes me as all-encompassing, a phenomenon that covers the whole range of human activities; whereas Aron tends to pinpoint only certain aspects, especially in his studies on industry.

In my research on technique, I was ultimately led to situate modern technique in relation to the past. This is obviously very much on the minds of those who say to me, "But people have always used techniques." Of course, people have always used techniques; nor can we say that what we are now doing is unrelated to what was done in earlier times. Nevertheless, I feel we should not reassure ourselves by saying it's basically the same thing. According to Emmanuel Mounier (and this is one of the reasons I broke with him), there is only a difference of degree between a flint arrowhead and the atomic bomb. In this case, I would have to very firmly apply Marx's notion that, on a certain level, quantitative change is qualitative change. Hence, when the human race moved from the flint arrowhead to the atomic bomb, there was a *qualitative* change. Mounier also said: "When you admire techniques so much, just look at your own hand. Is there any technical device more perfect than your own hand?" Well, that's true, of course. But I don't think that this notion allows us to understand in any way the singular and unique character of our age.

In other words, I was led to distinguish between what people were doing in all other societies when using certain techniques, certain technical operations. Clearly, any action of hunting, fishing, building a cabin, even gathering, is a technical operation—a practice. On the other hand there is the phenomenon that we have known in the Western world since the eighteenth century and that has developed during the nineteenth and twentieth centuries; I call it the technical phenomenon.

The great difference between the two is in their respective characters. First of all, there is the participation of the rational. Until the eighteenth century, technique was, purely and simply, a practical matter. In the eighteenth century, people began to think about techniques: they compared them and tried to rationalize their application, which completely changed the perspective. A technique was no longer merely a practice, it was no longer merely an operation. Now, technique passed through a rational intervention, and it had a completely different object; its object was efficiency. When studying the old techniques, one is extremely surprised to see how unimportant efficiency was as a decisive or determining notion. Techniques were used for religious reasons, for purely traditional reasons, and the like. If one technique were more efficient than another, that didn't trouble the users very much. The technical phenomenon, however, is characterized by evaluations of techniques, and comparisons in terms of this criterion of efficiency.

Hence, the technique existing in the Western world since the eighteenth century is qualitatively different. This is not only a question of volume. Technique has assumed different functions. This is the second element which differentiates pre- and post-eighteenth-century technique; technique has left the framework of material applications. When speaking about technique, we have always habitually thought of the machine. But I feel it is a grave mistake to regard technique essentially as machines.

At the present, with the development of information techniques and communications techniques, people are coming to realize that the machine, although not a secondary phenomenon, is certainly one of many phenomena in technique. Research on rational and efficient methods is expressed not only in constructions of material devices— machines. It covers and has gradually come to encompass all human activities.

By this, I meant that there is now a precise knowledge of how a group or a society is constituted, evolves, and how one can organize to achieve a certain result. Sociology and psycho-sociology supply us with means to obtain the best returns from a work team, to "place" individuals in a given spot at a meeting in order to increase or decrease their influence, to make an organigram of an organization so that it will be as efficient as possible, to know whether it is better to establish long-distance or short-distance relationships in an administration, and so on. These are simple examples of what I mean when I speak of the techniques of organization in a society. They have been widely applied in human relations, public relations, and the army.

Psychological techniques are exactly the same thing. For instance, I have studied propaganda techniques and advertising techniques, and these *are* techniques. Hence, we see that the technical phenomenon covers not just a small part of our activities—those in which, as is often said, our muscular activity is replaced by the machine. The technical phenomenon is tending more and more to encompass *all* our activities.

There are techniques that we obviously are well acquainted with. Anyone who is involved even slightly in athletics knows that they are no longer left to the intuition of the athlete; today, they are extremely rigorous techniques. A century ago, sports were very spontaneous. Runners or swimmers each had a "style," and each improved individually. But since then, more and more precise rules have been established. A champion's life is thoroughly programmed (food, sleep, and diversity of physical training). And people have minutely studied (often on film) every single gesture, pointing out an error here, a slowdown or speedup there, endlessly correcting each movement so that utmost efficiency may be achieved. Likewise, people have set up a "strategy," seeking the right moment for the runner to accelerate to the maximum. . . . All this is technique.

We are dealing with what is basically a power covering the full range of human life. This expansion of technique to human groups, to human life, is one of the essential characteristics of our world.

A last crucial feature, it seems to me, is the relationship between technique and science. Here too, people normally view technique as an application of scientific discoveries. But this schema is far too simple. At the present, we are faced with a highly complex and ambiguous situation; science can evolve only with the help of technique. One need merely recall the exploration of space to see that science is now tied to the information that the many techniques contribute. In other words, there is no linear relationship between science and technique. The relationship is, first of all, mutual: science/technique and then technique/science.

Beyond that, however, technique likewise results from its own conditioning. A technical innovation is not necessarily the fruit of a new scientific discovery, but most often is an internal, intrinsic development of technique itself. This means that we no longer need science in order to combine several techniques belonging to different domains. These techniques interconnect and combine, resulting in something new, something technically new. Likewise, we know how sterile some scientific discoveries can become for various reasons, over a long period of time, and never flow into the technical domain. Hence, we must abandon this simple view of the relationship between science and technique.

This analysis of the technical phenomenon, along with other factors, led me to criticize the current analysis of the Industrial Revolution. I felt that scholars were overemphasizing the purely industrial phenomenon. The technical revolution, I believed, had already been launched, and the Industrial Revolution was only one of many aspects of it. What permits me to say this is the observation of what took place in Western society when the Industrial Revolution was developing. The state, let me note, appeared at the same time, and in the modern state, with all its structures, one can also note the emergence of administration with a trend toward administrative efficiency, rationality, the use of completely modern devices. We can see the same tendency toward rationalization in law, and we must also recall the rationalization of science, which, having progressed slowly during the fifteen and sixteen hundreds, was truly rationalized in the eighteenth century.

In other words, by taking these three examples—the state, the law, and science—I perceived that it was not only in the area of industry that the technical mentality emerged, along with the concern for rationality and efficiency; this development occurred in many other domains as well. Thus, the great phenomenon during the eighteenth century in Europe was not the use of coal and the construction of machines. It was the change in the whole society's attitude toward a new fact: technical practice. The Industrial Revolution was just *one* aspect of this new practice.

It is astonishing to see historians misinterpret in this way. One need only consult Diderot's *Encyclopédie* in the eighteenth century to realize that people were fully conscious of this change in attitude. At that time, there was enormous interest in machines, but machines as one aspect of technical innovation, as one aspect of the new understanding of human beings or the new understanding of society, which now had to be rational and efficient. One finds this new conception throughout the *Encyclopédie*. Thus, if one no longer regards the Industrial Revolution as the dominant element, the determining element, the problem becomes far more vast and complex. There is no longer just an economic problem, say, of how people passed from the craft stage to the industrial stage. The problem is now, why do people apply certain processes both in industry and elsewhere—processes that might have been known in the past but that had never been applied?

It seems to me that certain conditions that had never existed before came about in the eighteenth century. I might very summarily indicate five of them, which, simultaneously, allowed the development of the technical phenomenon.

The first was a significant growth in population. This increase presupposed a better organization, but also the availability of a work force, as well as a far denser, far more dramatic circulation, not only of people but also of ideas.

A second fact was what I might call the social plasticity. That is to say, a very large number of *ancien régime* social structures were destroyed in England and France and then in Germany. Social groups crumbled, and their members had the possibility of moving toward completely different activities. This development turned out to be essential for the Industrial Revolution in the strict sense of the term; it also created a working-class population.

On the other hand (and this is the third aspect), the new era brought inventions by intellectuals and practitioners of a clear technical intention. They felt that one must be able to apply the same system of processes in all domains. This was an intellectual innovation.

And then (and this is the fourth element), this development was grafted onto a very long technical maturation, which went on for something like two hundred and fifty years in Europe. This maturation consisted of very small progressions which slowly accumulated, though never appearing to be decisive or to have any structure. This was contrary to what had occurred in the Roman Empire or in China during periods of technical developments. In addition, this maturation may also explain the emergence of the clear technical intention.

Finally, an important factor was the accumulation of capital for utilizing the industrial means, as well as all the technical means. Naturally, capital was necessary, especially when the private entrepreneur was operating. This was the first time since Roman days that Western society accumulated a certain amount of capital from commerce per se.

These five elements together led to the development of the old to the new society, one aspect of which was the Industrial Revolution.

In these conditions, technique, I felt, had gradually become the key phenomenon of our whole society, not only because it gradually encompassed all activities, but also because it could evolve only on the basis of certain values. That is to say, technique is not just a practice; it also presupposes values—an intellectual or a spiritual attitude consistent with the demands of technique. Furthermore, it requires a certain social structure. I just mentioned that the Industrial Revolution came about only because of new values—rationality and efficiency—and because of a change in social structures. Well, *what had occasioned the technical phenomenon now became a demand of technique for continuing its own development.* You see, in growing, technique requires that human values be in exact accordance with technical development and that social structures develop purely in terms of technique. This, I believe, shows that nothing in a society remains intact once technique begins to penetrate.

I should indicate that values which are indispensable to technique include utility values and, until very recently, work values. We must not forget that in ancient societies work was not a value. It became a value precisely when the techniques required that people be put to work. We are dealing here with a frequent misunderstanding. People always claim that techniques economize on work (and this is quite correct). *But this is based on the conviction that we are meant to work all the time!* This was by no means the conviction in earlier societies. For two centuries now, we must note, the West has worked a lot more than any previous society.

In reality, work has changed character. It is no longer a curse as in the Middle Ages. On the contrary, it has become a positive value because it is indispensable for capitalist and industrial development, and also for all technical development. All people must be integrated in the work process, albeit, of course, with the hope, with the promise, with the utopian expectation, that we will finally no longer *have* to work! This is part of the dual effect of technique, which makes people work to their maximum (Taylorism—time and motion study—is one aspect of technique), but always with the prospect that technique will totally and radically take over for us and replace us when we can finally do nothing. In the nineteenth century, this became an essential value of a world dominated by technique.

There is another essential value, however, and that is happiness. I would like to cite what Saint-Just said: "Happiness is a new idea in the world." He was right. Happiness was indeed a new idea, but not in the elementary sense that other societies had never had the notion of happiness, or that people had never desired happiness. The new element was that people now realized that happiness was based on certain material conditions. The eighteenth and nineteenth centuries abandoned the idea of spiritual or intellectual happiness in order to have this material happiness, consisting of a certain number of essential consumer goods. And hence, in the nineteenth century, happiness was linked to a well-being obtained by mechanical means, industrial means, production. The new thing that Saint-Just spoke about was that, in the past, happiness could appear as a very vague, very distant prospect for humanity, whereas now, people seemed to be within reach of the concrete, material possibility of attaining it. That was why happiness was to become an absolutely essential image for the nineteenth-century bourgeoisie, and for modern society. Happiness was attainable thanks to industrial development, and this image of happiness brought us fully into the consumer society.

Now one can almost say we have come to realize that consumption does not assure happiness. We are passing through a crisis, a crisis of values. I just mentioned that work too, as a value, was passing through a crisis. It was the new development of technique that brought about a crisis in the values that allowed the initial development of technique. I also noted that technique not only presupposes adapted values but also demands a social structure allowing the development of technique. We must realize one very simple thing. Every time technique penetrates an environment that is not made for it, it will upset that environment. I am thinking especially of how technique and the techniques involved in industrialization are influencing the Third World.

We can say that wherever the local work force was called in for purposes of industrialization, the result was total disruption—not just partial, but total disruption of the entire country. The reason is very simple: the people who become workers in industry leave their families and come to the city. Not only do they work at jobs previously unknown to them; not only do they earn their livelihoods in a different way; but, above all, they completely escape the social control of their milieu. They now live in the city, uprooted. They have escaped the natural authority of the paterfamilias, and their resources no longer depend on the tribal or patriarchal structure. They have their own individual resources. In other words, the mere summoning of workers causes a destructuring of the family, a setback and ebbing of the economic mode in which the entire population lived, and a certain moral uprooting.

Perhaps we should expand on this point. Traditional societies, we must recall, have no individual morality. Indeed, morality is really the normalized behaviour of the group, with each person individually expected to live as the group does. Once people are torn away from the group and live as workers in industry, then, whatever their level, they no longer depend on the social control of the group. They then need an individual morality to compensate; but they have none.

These people have not gone through the long process. They have not travelled the long road which took centuries in Europe: the long transition from a tribal structure to an individual morality. Hence, the disintegration we perceive wherever an industrial development begins in a Third-World country.

From a social point of view, however, I think that we can also note the transformation of our own society under the impact of technique. And here, I would like to indicate the difference between our society dominated by technique and the societies issuing from the Industrial Revolution—that is, between mid-twentieth-century and nineteenth-century society.

First of all, we have witnessed the appearance of a new class, a new ruling class. Marx was perfectly correct in his analysis of the role of the ruling class, which was the role of the capitalists. The capitalists held the power because they held the economic instrument on which everything depended. But now we see a new ruling class emerging, the class of technical experts, which represents one of the real aristocracies in all our societies. Many sociologists (who by no means have the same perspectives as I) have established the banal formula that in our society success depends not on *what you have,* but on *what you know.* It is more important to be competent, to be a high-ranking technical expert. This assures you a far more important career in society than starting out with a small amount of capital, which may perhaps allow you to set up a small business, but will not really allow you to make it in our society.

In other words, the person who has knowledge—practical know-how, technical know-how—is the true master in society. At the present, if one's capital is not put to work by people with technical know-how, then it will not count. The person owning capital privately is becoming less and less important, compared with the person who activates his or her capital within the ensembles of technical operations. And this class is the ruling class in that, like all traditional ruling classes, it possesses certain secrets. The technical expert's knowledge is always a mystery to non-experts.

Are we therefore living in what has often been called a "technocracy"? I do not think so. This is, I believe, a misuse of the word. In no society do the technical experts exercise complete political power such as is exercised in a democracy, an aristocracy, a monarchy, and so forth. No, the technicians do not hold the power. However, a certain trend toward technocracy is apparent. For instance, in the Soviet Union, it is more and more the technical experts who directly exercise the power. And this is a question constantly asked in France, for example, in regard to the president of the Republic. Ultimately, the development is toward groups of experts in the most rigorous sense of the word. Although not truly a technocracy, this is nevertheless an aristocracy. And that is why our societies, whether Socialist or capitalist, boil down to exactly the same thing. Our societies are aristocratic societies. Here, I would like to cite an excellent study by the Yugoslav Milovan Djilas on the new class. He was one of the first to perceive (and others followed suit) that the Socialist world also had a new class division. This division is no longer between the owners of capital and the proletariat, but between those who control the bureaucratic, administrative, scientific, and other techniques, and those who do not control them. The former group is truly a new class. Meanwhile, as this new class emerges in our society, we note a trend toward a diminishing opposition between the former bourgeoisie and the working class.

The explanation for this phenomenon is long and difficult. I have just said that the classical bourgeoisie, the bourgeoisie of independent means, has disappeared. The middle class has now moved toward technical functions; and in the working-class world, there have certainly been ruptures. One can no longer compare a longshoreman's condition to that of a highly qualified worker who is actually a technician. However, Alain Touraine, a French sociologist, has observed a significant difference between them; a worker who is only a practitioner can have an excellent practical knowledge of techniques, but he or she will never reach a superior level in society, because only a theoretically trained technical expert can mount that high. Technique must now be known not on the level of its practice, but on the level of its scientific foundations. As long as one has not made this transition, the limitations of improvement on a practical level are quickly reached. Real changes are now only made on a theoretical level by means of a science-based technique.

In other words, we see that technique is modifying the structure of our entire society. We are thus dealing with a phenomenon that not only changes our habits—we fly planes, watch television—but also ultimately changes our political interpretation. Certain parties still deploy their action, and all their propaganda, as if the situation was still one of a dominated class, a poor class, and so on. They thereby preserve Marx's nineteenth-century interpretation of the opposition between the bourgeois class and the working class. But this traditionalism is almost a century behind the times.

I am not saying that there are no more wretched people, I am not saying that there are no more dominated people. There certainly are. But now, power is no longer in the same place. Power is no longer in the hands of the owners of capital. I can develop this idea by analyzing the multinational corporations. Here, as we clearly see, capital still exists; but it is now structured in terms of technical demands rather than in terms of the ideas formulated by a capitalist. Today, there is no longer any owner of capital who plays the part that could once be played by a captain of industry.

Technique thus now appears as both a key phenomenon and as a point of view. I should elucidate these two levels and the difference between them.

Technique is a key phenomenon. In other words, for me, it is a reality, it truly exists. When I speak about technique as I do, I try to present what I perceive. And on the basis of hundreds of observations, I can study technique as a key phenomenon. But at the same time, technique is an instrument of knowledge, a scientific instrument. It offers the central viewpoint in which one must place oneself if one wishes to understand and explain what is happening. Hence, there is a double element: the epistemological element and the reality element. All phenomena in our society are either an imitation of technique or a compensation for the impact of technique. These, I believe, are the terms in which to analyze most of the realities of our world. By imitation, I mean under the immediate influence of technique, which directly moulds, for example, the administrative system. When the computer enters administrative practice, we must refashion administration to imitate the computer. Hence directly consequent and imitative mechanisms.

At the same time, however, the refashioning necessitates mechanisms of *compensation,* because it is extremely difficult to live in a technical universe. Just think of the countless science fiction stories. The technical universe, which ought to be a rational universe, is an extraordinarily icy, extraordinarily alien universe. People cannot be happy in a purely technical milieu. They can no more live spontaneously in the technical milieu than the astronaut in the cosmos. The astronaut must be powerfully equipped for survival in a space environment. Likewise, a person, no matter who, cannot live totally in this rigid, rational, icy world that is the world of technique. I am not saying it will always be impossible. After all, perhaps people will adjust to a rigid, rational, and icy universe. But for the moment, they have not adjusted. For the moment, a human being is still an extraordinarily irrational creature. It was a tragic error of the eighteenth and nineteenth centuries to believe that people were originally rational beings and that all irrationality must be suppressed. Each person is a creature of passions, of flesh and blood, a creature of impulses and desires. Hence, when a person lives in a purely rational framework, it is impossible to be happy. He or she then requires compensations; and a very large number of factors characterizing the modern world are purely compensatory factors, making up for the impact of technique. We have no choice but to live in this world dominated by technique; but we are forced to find something providing satisfactions elsewhere and permitting us to live otherwise. This state of affairs is felt very deeply, especially by the young, because technique has two consequences which strike me as the most profound in our time. I call them the suppression of the subject and the suppression of meaning.

First of all, the suppression of the subject. Technique is an objectifying power. If a person has learned to drive a car correctly, then it doesn't matter who he or she is, it's all the same. The subject, if you like, cannot indulge in purely subjective fantasies in a technical framework, but must act as technique demands for that one act. The suppression of the subject is transforming traditional human relations, which require the voice, which require seeing, or which require a physical relationship between one human being and the next. The result is the distant relationship. If we compare the countless telephone calls we receive throughout the day with the personal relationship we have with one or two people, we realize that our distant relationships are considerably more numerous. And in distant relations, there is really no subject. Technique brings about the suppression of the subject. This result is accepted by a certain number of intellectuals in France such as Michel Foucault, who feels that one can very easily abandon the subject. And yet Foucault has not stopped using the first-person pronoun. He still says "I." That is, willing or not, he considers himself a subject. He says "I do" or "I think." This is not "one thinks." This is not just anyone or anything.

In other words, while technique leads to suppressing the subject, we do not experience it at all well. We still feel we are subjects, we still want a very personal, unique encounter. Hence, we are in contradiction with the technical milieu.

Then, there is the suppression of meaning; the ends of existence gradually seem to be effaced by the predominance of means. Technique is the extreme development of means. Everything in the world dominated by technique is a means and only a means, while the ends have practically disappeared. Technique does not develop toward attaining something. It develops *because* the world of means has developed, and we are witnessing an extremely rapid causal growth. At the same time, there is a suppression of meaning, the meaning of existence, the meaning of "why I am alive," as technique so vastly develops its power.

We know *that power always destroys values and meaning.* Here I would point out remarkable studies done by Friedrich Junger on the conflict between power and meaning. Wherever power augments indefinitely, there is less and less meaning. One seeks a meaning when power allows us to be *ourselves*, without being superhuman. Thus, we have these two extremely active factors—the suppression of the subject and the suppression of meaning—both due to technique and both making humanity very uneasy and very unhappy.

I will try to show this interplay of imitations and compensations in certain areas, for instance in art or religion, or perhaps politics. I could start with the example of politics as a field in which technique has completely transformed the conditions of power. What strikes me is, on the one hand, the extraordinary increase in the means of action by the state and, concomitantly, the stunning decrease of power by the individual politician.

The modern state has means that are all technical: administrative means, communications means, control means, means for planning land use, all the information means that no other state has had until now. Hence, we are dealing with a phenomenon that is very different from the one studied by Weber. Weber did see the growth of the state. But nevertheless, for him, the state was always tied to the power of a certain category of politicians. Bismarck, I would say, was ultimately the model of the state. But now we no longer need a Bismarck, we no longer need great statesmen. For with this augmentation of means, we witness the lessening importance of the politician.

The politician is someone who is not a technician, who does not know the means that the state can employ, who depends in all decisions on what the technical experts say and on what those other technical experts, the bureaucrats, do. Every politician must first deal with dossiers prepared by groups of technical experts, and these dossiers contain a decision ultimately suggested by technical experts. This is the decision that the politician will always make. Once the decision is made, it has to be applied by other technical experts, the administrators. The politician now has a tiny role, especially since the administrators—without even saying anything—can block this decision, so that nothing will happen. We can see this very concretely in France with the president of the Republic, who has generous ideas. He launches certain proposals, but nothing happens.

When I describe the increase in the means of the state and the decrease in the power of the politician, I am speaking of something that has been taking place in both the Socialist and the capitalist world. Indeed, this is one of the points of progressive identification between the two. Within this framework, one can say, for instance, that law is totally losing its validity and significance. It is becoming a technical device for administration and organization. In other words, law no longer has the objective of bringing justice. Today it is an instrument in the hands of administrators, in the hands of the state— an abstract instrument for administering and organizing the society.

We often have the situation (not only in France) of administrators acting outside of any legal rule and then, after acting, making juridical decisions that simply legitimize what they have done. In other words, law is no longer made in advance for the administrator to obey and apply. Law is made *after* the fact to justify what has been carried out. Here, we are patently faced with the loss of meaning—law no longer has meaning—and along with this occurs the disappearance of the subject—for the politician was once the subject. Through two centuries of European history, great individuals forged all history. Today this is no longer true. History is made by the heavy mechanisms of the state machinery and by the social forces that combine with or contradict one another—hence, things that totally escape the power of the subject.

In a word, we are witnessing an imitation of technique by the bureaucratic and technocratic power of the state; and we are witnessing a compensation, I might say, by the *discourse* of the politician. The speeches of the politician are always very important, indeed enthralling, because we think we are in charge of the situation. When we listen to a politician, we agree, we disagree, we contest what is said. But instead of looking at the reality of what is occurring, we are content to have a person on stage who tells us: "I am in charge of the situation," or we argue that he is not in charge of the situation. Either way we feel we have a subject before us and we feel that we are subjects. That is how we make up for the absence of the politician's power and our own power. In fact, in regard to technical growth, politicians are utterly devoid of means. They simply cannot reorient our society in a different direction.

That is one brief example. A second example will show what I mean by the disappearance of meaning and of the subject as well as the double movement of imitation and compensation in the domain of art.

Modern art is completely characteristic of this influence, this impact of technique. It is characteristic not only in what the artists produce, but also in the explanations offered by critics for modern art. There is no more subject. We know all the theories on painting. Now we have forms, we have splotches of colour. But this means nothing. There is no theme.

The same holds for the novel. In what was known in France as the *nouveau roman*, the subject was suppressed in the following sense. There was no plot. It was considered totally retrograde to tell a story in a novel; you did not tell a story anymore, and there were no characters either. So we had those utterly amazing novels in which there was never anything but "I, you, he," and we never knew who "I" was, who "you" were, who "he" was. There was total confusion sentence by sentence, and there was no telling what referred to whom. Hence, we were dealing with the expression of an art that reflected technique, the suppression of the subject.

A book of thoughts on modern art, explaining how to use the computer to paint a picture, showed a reproduction of a painting done by a computer and a painting done by Kandinsky. The author asked: "Who is the artist here, the computer or Kandinsky?" It was impossible to answer one way or the other; and so, it was said, you can see that the computer can do real and authentic painting. I, however, would say that this simply means that Kandinsky paints like a computer. That's all. It does not mean that the computer paints like Kandinsky. In other words, the painter has taken lessons from technique, he has taken lessons from the instrument, and he reproduces by suppressing the subject.

The same holds true when someone makes music with the computer or when the musician proceeds not by listening to sounds, by creating on an instrument, but by a mathematical development which is translated into certain sounds. This process is totally different from what artistic creation used to be. Now, we truly have the reproduction of technique by art.

Likewise, the suppression of meaning. How many times have we not read, particularly in linguistics and above all in structural linguistics, that we must never seek the meaning of a text. A text simply exists. There are black forms on white paper, and we have to read the text as it is. We have to see the structure of the text, and it makes no difference whatsoever whether it has any meaning or not. Remarkably enough, there is a whole category of artists and intellectuals who fully accept that language has no meaning, that it simply has structures. In a very recent article, a modern linguist actually stated: "Naturally, when we say 'Please pass the bread,' this sentence has meaning. But this is quite unimportant because it is only an *extraordinarily rare case of language.*" For my part, however, I think that this is the *habitual use* of language; the structural utilization of language in modern poetry does not strike me as the normal and habitual case. Again this example illustrates the suppression of meaning, and here too, a tendency to imitate what is happening in technique.

But there is also *compensation.* It is not possible to live only in icy painting, in abstractness. It is not possible to listen only to computer music. So, we let off steam in the opposite direction; we dash toward, say, Pop Art, to make up for the technical milieu we live in. We move toward total sexual liberation as a compensation. Erotic spectacles make up for the far too sophisticated technical spectacle. Hence, modern art, in the suppression of subject and meaning, has two directions. It is a pure and simple reproduction of techniques; and it is compensation for technique itself.

My final example of compensation involves nearly all the religious phenomena in the present-day world. We know that there has been a sudden development of religious phenomena. Personally, I do not believe that this development comes (if one is Christian) from the Holy Spirit. It is quite comprehensible, from a purely sociological viewpoint, in the light of technique. Life in our technique-dominated world is extremely frustrating and extremely distressing, so we have to escape it. Religion appears as a means of escape. That is why religion is taking the forms that we now witness, the forms of spiritualism, and the extremely ardent, extremely intense sects of pietism, through which people can separate themselves totally from the world. Technique is coming to dominate the material world, and we are subject to the material world. But we can compensate by way of religious escape, by way of spiritual escape.

We must not forget that this is what Marx meant when he called religion the opiate of the people, when he said that the function of religion is to continue the domination of capitalism over the exploited and to make them believe that in paradise they will have freedom and no longer be exploited. Now, of course, the characters have changed; it is no longer the capitalists and the workers. The phenomenon is completely different, and more abstract. We now have technical organization on one side and human beings, all humanity, on the other side. Religion plays the same role here, allowing us to escape, and to continue living at the same time. Clearly, what is happening in Iran is a compensation for an overly sudden and overly rapid technicization by the Shah. The people could not bear this sudden transformation and have therefore fled into mysticism.

I would like to specify that all the things I have discussed in relation to technique are not primarily theoretical. They are, essentially, observations of what I see as the substratum or, in Marxist terms, the infrastructure of our society. On the basis of these perceptions, I have developed an overall interpretation by means of a theoretical effort. I have set up a theory, which in my eyes, however, is nothing but a formulation and an account of what I have observed. In no way is this a closed system. I am obliged to take heed of any new fact that I note, and I must then change some element in the original construction. (Here, once again, I think I am being quite faithful to the ideas of Marx, who kept rethinking his theoretical givens in line with economic or political events, for instance the Commune of 1871.)

It is obvious, for instance, that the events of 1968 and the development of the hippy movement led me to revise a certain number of conclusions I had drawn about the effects of technique on humanity. I was, I might say, more pessimistic before 1968 than after. I used to think that we were so trapped in the technical system that we had no further resources to draw on. And then 1968 brought an explosion which opened certain paths and which showed that we were not truly conditioned.

By the same token, the religious movement in the Soviet Union, which is very different from the religious movement in Europe and America, shows that people have been psychologically conditioned, whatever technical methods may have been employed to shape them. This led me to modify a certain number of my judgments.

A further new phenomenon which is equally essential is, of course, the spread of the computer. So long as the computer was a very particular, very piecemeal phenomenon, it could not be a focal point in a study of society dominated by technique. But now I must rethink a good portion of my theory of the world dominated by technique because the computer is having ubiquitous consequences unlike those of any previous technique. In other words, my theory is open-ended. The computer has always been a part of the world dominated by technique, but its extensive application has altered the functioning of this world, and this is something I began analyzing several years ago.

Of course, mine is a general theory in that it allows me to interpret certain facts. I would say that the more facts a theory takes into account, the more valid it is. Heeding as meticulously as possible everything that occurs in our world, my theory of technique, my analysis of technique as a system contributes to understanding more facts, I feel, than most other present-day theories that I know of, including classical Marxism, which is obliged to place most modern phenomena in parentheses. My theory is a means of interpretation, which strikes me as all the more serious in that I am not obliged to modify the facts in order to maintain my doctrine. In reality, the theory I have constructed allows me to verify a large number of facts. To the extent that it is evolutive in itself, I think that I can integrate more and more facts.

Finally, I should state that I have not offered a metaphysical theory or a metaphysical system. I have remained solely on the level of the reality occurring in the present world.

3

The Present and the Future

A CERTAIN NUMBER OF French sociologists, particularly Georges Friedmann, believe that technique is a new milieu in which people live. But generally, none of them has analyzed this phenomenon or drawn all the inferences from this observation. I have developed this problem of technique as a milieu, and I have interpreted it first on the basis of the experience of the environment in which we live. Principally, this milieu is the city, an entirely artificial world. There is practically no living element here except for the human being. The city is a pure product of techniques of all kinds. In short, we live in a milieu that is totally dead; one of glass, steel, cement, and concrete in which technical products replace the old natural milieu in which we used to live.

This urban people's contact with nature is totally accidental and frequently very slight, for instance when they go on vacation. But when they do take a holiday, they also wish to preserve the artificial milieu. We see vacationers—at least in France—surrounded by countless gadgets such as TV and radio sets; even when in contact with nature, they need to reconstitute a technical milieu.

However, we should not deceive ourselves about the meaning of this milieu. Just what is a milieu? It seems to me (and there are few studies on this subject) that it is not only the place in which a person lives, but also the place from which a means of survival is drawn. Of course, this is extremely simple. But at the same time, the milieu is *what puts one in danger*. Hence, a milieu both makes living possible and also *forces change*, obliges us to transform who we are because of problems arising from the milieu itself.

In other words, I feel there is never a true and total adaptation of the living creature to the milieu. There are successive nonadaptations with challenges, of course, and then new adaptations, hence changes. This permits me to define the milieu for all living beings. We know, of course, that some animal species have vanished for failing to adjust to a changed milieu. Although initially adapting, they ceased to do so.

We have succeeded in overcoming various crises caused by the milieu we live in. Hence, we must pinpoint these two elements. The milieu is that which offers the means to live and also that which poses problems and dangers.

There are rather fundamental consequences of this transformation of the milieu in which we have always lived into a milieu of technique. They are fundamental, I think, in that we must call upon a theory which is entirely new, the theory of the three milieus. For it is not true that we have passed directly from the natural milieu to the technical milieu. In reality, we have known not two, but three successive milieus: the natural milieu; the milieu of society; and now, the milieu of technique.

The natural milieu was that of the prehistoric period, when there was no organized society as yet and when immediate contact with nature was absolutely permanent. This was really an immediate contact; nothing mediated, nothing served as an intermediary between the human group and nature in the traditional sense of the term. Nature provided a sustenance for human beings, who lived by hunting, by gathering; and nature also provided their principal danger—the danger of poisons, the danger of wild animals, certainly, but also the danger of barrenness, the danger of shortages. This was the first milieu, the one we think of quite spontaneously.

However, humans found a way of defending themselves against this natural milieu, getting the best from it and protecting themselves against it—something that would mediate between themselves and nature. This new means was society. The creation of human society appeared with the times traditionally known as historical. History is tied not to the existence of a natural milieu, but to the existence of a social environment. Society allowed humans to grow strong. The human group became an organized group, a group that has gradually dominated the natural environment, using it as best it can.

What I call the "social period" is the historical period beginning some seven thousand years ago, when human beings succeeded in more or less protecting themselves against nature and taming it, in grouping into societies and in utilizing techniques. During this period, society was the natural milieu for human beings, who remained in close contact with nature (there was a balance between town and country). Techniques were only means, instruments. They were not all-invasive. The great problems were those in the organization of society, the political form to choose, the distribution of labour and wealth, the circulation of information, and the maintenance of cohesion among groups. Thus, society was the environment which allowed human beings to live, and also caused problems.

But while becoming a human milieu, society also turned into something that allowed us to live and then imperilled us. For the chief dangers were now wars, which are an invention of societies. The social milieu still seems like a "natural milieu," because people to some extent remained in nature. Throughout the historical period, this social milieu marks the intermediary period between the natural milieu and the one we know today, the milieu of technique.

The third milieu, this technical one, has actually replaced society. Not only are natural data and natural facts utilized by technique, mediated by technique; not only are people alienated from nature by technique; but also social relations are mediated and shaped by technique. In short, the weight of society is far lighter now than the weight of technique.

Of course, when I speak of these three successive milieus for humanity, I am certainly not saying that the appearance of a new milieu eliminates and destroys the preceding one. I have just mentioned that when human beings organize themselves into a society, they still remain in contact with the natural milieu. Society is a means for best utilizing the means of nature and avoiding the disadvantages of the natural environment. By the same token, it is obvious that technique does not suppress nature or society; rather, it mediates them.

Nature was mediated by society, with people living in the social group and beyond nature. Now, technique mediates society and, on a secondary level, nature.

Each preexisting element—nature or society—is to some extent obsolete. But it still exists in regard to dangers. For instance, the dangers of natural epidemics were always imminent in the social milieu. However, epidemics were a relatively less serious matter than the dangers inherent in society. Likewise, there are natural dangers and societal dangers that still survive, even though the milieu we now live in is a technical one. There are still typhoons and earthquakes; there are still wars and dictatorships. Yet in reality, all things are already rendered obsolete and placed on a secondary level by the emergence of a new milieu. In other words, the problems raised by a former, obsolete milieu are no longer the essential or fundamental problems.

When human beings were organized in a society, their fundamental problems—and this was the whole question of politics—were the very organization of society, the relations between various societies, the growth of political power, and the control of political power. These issues were far more important than those concerning natural phenomena.

Similarly, today, the technical phenomenon, including both the positive and the negative aspects of technique, the things that both endanger us and increase our power, are far more important than the problems caused by society itself. Hence, we ultimately come to the following conclusion: most of the problems we face today—especially the purely political ones, which relate to the foregoing historical period when the essential milieu was society—are all obsolete by now. These are ancient problems, if you will. During the historical period, it was more important to solve political problems than a certain number of purely natural problems. Likewise, today, it is more important, more decisive, to solve the difficulties raised by technique, the dangers coming from technique, than to solve purely political issues, the problems of elections, the question of whether a system should be democratic or not.

Of course, just as society employed the means of nature, so too technique employs the means of society. Hence, technique aggravates political problems. Political power is now in the hands of technical structures that far surpass any power ever held by older political authorities. However, this is no longer a political problem. Whatever the regime, it has its structures in hand. The problem is actually a technical one.

Thus, technique has become a milieu. Beyond that, however, it has also become a system. I am using the term "system" in a sense that has now become customary since Ludwig von Berthalanffy: an ensemble of mutually integrated elements, situated in terms of one another and reacting to one another. On the one hand, every element in the system is understood only in terms of the whole, in terms of the system. Any variation in the whole has consequences for the integrated parts. And reciprocally, any change in the elements affects the whole.

This, I feel, is a new view of technique, with a difficulty that I have already pointed out: when I speak of technique as a system, I mean two different things. The first is that technique has in fact become a system. This means that each individual technique is actually integrated in a totality; each datum of technique must be understood in terms of this totality. Hence, there is actually a system of many techniques. Secondly, when I say that technique is a system, I mean that the concept of "system," used both philosophically and sociologically, is a means of interpreting what is happening technically. It is essentially an epistemological instrument allowing us to know and understand technique better. Hence, the term "system" designates both the fact and the instrument of comprehension.

This interpretation of technique as a system has enormous consequences. I will mention only two.

First of all, technique as a system obeys its own law, its own logic. In other words, we are dealing with an autonomy of technique, a closure of technique in itself. There is a very small margin of possibility for intervention, for outside action—economic, political, or whatever—on technique. Furthermore, technique is autonomous in regard to morality, politics, and so on.

On the other hand, it is involved in a process of self-augmentation. Technique augments itself for its own reasons and with its own causalities. We would obviously have to go into a long explanation of how the person who interferes with the milieu of technique and the system of technique intervenes to some extent as an instrument of technique and not as its master. Technique has the power of self-augmentation, which is intrinsic to it.

We encounter an apparent difficulty here. The system of technique progresses by virtue of its intrinsic laws, and there is an autonomous process of organization. But at the same time, this can occur only by means of constant human decisions and interventions. By describing the system as autonomous, I do not mean an autonomy capable of directing itself and reproducing itself without human intervention. What happens is that the system determines the one who must make the decisions and who must act. The *sole* actions and decisions to be allowed are the ones that promote the growth of technique. The rest are rejected and quickly forgotten. Those who make the decisions are neither aesthetes, skeptics, critics, nor people free of obligation. Since childhood, they have been accustomed to technique: they feel that only technique is important, that only progressive thinking is valid; and they have learned techniques for their work and their leisure. In this way their decisions always support the autonomy of technique.

Here we have a problem. Like any system, technique ought to have its self-regulation, its feedback. Yet it has nothing of the sort. For instance, if one observes a set of negative effects caused by a group of technicians, one should not only repair the damage, but go back to the origin of the techniques involved and modify their application *at the source*—for fertilizers, say, or certain work methods, or chemical products. But this action is *never* taken. We prefer to let the drawbacks and dangers develop (on the pretext that they are not fully demonstrated) and to create new techniques to "repair" the problems. In fact this actually entails a positive feedback. There is no self-regulation of any kind in the system of technique. This does not mean that it is not a system. It does, however, mean that we are confronted with a system that has gotten out of hand—a system incapable of controlling itself. Hence, we cannot expect any rationality, contrary to what we may believe. And this, I may say, is going to be the chief danger, the chief question when we think of ourselves within the system. That is a first set of consequences.

A second set of consequences is that, contrary to what we usually do, we can no longer understand technique per se. No technique can be understood in itself because it exists only in terms of the whole. Yet that is what we always do when, for instance, we consider television. We ask: "What are the effects of television? Can one escape the impact of television? Can we master television?" And the reaction is always totally elementary: "But I'm not the least bit addicted to television. I can switch off my set whenever I like. I'm completely free." We respond as if television were a separate phenomenon, likewise independent of the system. The same is true for the automobile. Observers are investigating, for example, the effects of the car on an individual or on an entire populace, as though the car were not located within the system of technique, a part of an extremely complex set of techniques.

However, if we wish to understand television, we have to place it within the system of technique, that is, television in relation to advertising, in relation to the fact that the world I live in is turning more and more into a visual world, or that I am constantly learning that only the image that I see corresponds to a reality or that the world in which I am likewise makes constant demands by way of a growing consumerism. This is the same world in which I am obliged by the group in which I live to keep up to date on whatever takes place. I am by no means free to watch my television set or not to watch it, because tomorrow morning the people I meet will talk to me about such and such a program, and I do not want to put myself on the fringes of the group.

Likewise, I am part of a world in which the technical operation requires a certain amount of knowledge. I cannot enter a milieu or a job if I do not possess a certain quantity of knowledge, and a good portion of this knowledge is transmitted to me by television. Hence, in reality, I am not independent of my television set. With the set belonging to me, I am integrated in a totality that is the society dominated by technique, of which television is a part, and I am absolutely not free in my choices, in my decisions.

Naturally, I can decide not to watch a certain movie or program. But am I really sure that I can decide? For I am also a person who spends my day working at a generally technical job that is quite uninteresting, repetitive, and anything but absorbing. In the evening, what do I have for relaxing, for relieving the buildup of nervous tension that I have experienced all day long? Television. Hence, in a sense I watch television as a reward at the end of the day, and this too is caused by my living in this milieu.

Therefore, I am absolutely not independent in regard to television; and it is no use trying to understand the effect of television as an isolated phenomenon. The true problem is the situation of human beings in the totality of the society dominated by technique.

I already mentioned the absence of regulation in the system. This non-self-regulation and another feature of technique, its ambivalence, prohibit any accurate forecasting of what may happen. We are always left with two hypotheses: Huxley's brave new world; or else the "disasters" foreseen by science fiction or the Club of Rome. Neither possibility is predictable.

In fact, Huxley's brave new world, where everything is normalized, is, as I will explain later, absolutely impossible. On the other hand, the disasters predicted by the Club of Rome strike me as equally improbable since all precise scientific forecasts about the world dominated by technique seem false to me. They are false because the system has no self-regulation, and we are incapable of foretelling the actual developments.

Then there is the ambivalence of technique, the fact that each emerging technique brings either positive effects or negative effects mixed in with the others. It is extremely simplistic and elementary to think that one can separate them, or to claim that one can suppress negative effects and retain the positive ones. Unhappily, this is never the case. I recall that when nuclear energy was launched, people simplistically said: "All we have to do is stop making atomic bombs and produce nuclear energy, and everything will be all right; we'll be pacifists." Alas, we know that the development of nuclear plants presents yet another danger and that ultimately every such plant is a potential atomic bomb.

Hence, the effects are by no means clearly separated. When we think of chemical products, we must bear in mind that a chemist comes up with a product of which we know certain effects. The secondary effects are only revealed a long time later; we are unable to discern them in advance. The same is true of fertilizers, medicines, and so on.

Thus, the positive effects and the negative effects of technique are closely, strangely interrelated. We may say that each technical advance increases both the positive and the negative effects, of which we generally know very little. I would therefore say that I cannot endorse either Huxley or the Club of Rome because of the margin of unpredictability. No scientific forecast seems certain to me. Nor can we now say that technique will keep progressing from innovation to innovation at the rate it has moved during the past thirty years, or that, on the contrary, we are veering toward a period of stoppage, of technical stagnation, which would obviously give us a certain amount of time, a delay. Clearly, a work like Huxley's or a cry of alarm like that of the Club of Rome is meant to alert us, to warn us of certain possibilities that lie ahead, but there is no way that we can tell which possibility is bound to come true.

Still, one thing seems absolutely certain: the difference and opposition between the development of the system of technique on the one hand and society and human beings on the other.

People have said, and I myself have written, that our society is a society dominated by technique. But this does not mean that it is entirely modelled on or entirely organized in terms of technique. What it does mean is that *technique is the dominant factor, the determining factor within society*, which is altogether different from Huxley's brave new world.

Society is made up of many different factors. There are economic factors, there are political factors. Human beings, as I have said, have an irrational element. Hence, being irrational and spontaneous, they are not fit for technique, and society, being habituated to ideologies, being historical and a result of the past, and existing in an emotional world of nationalisms, is as irrational as humanity and as unfit for technique.

The result is a shock, a contradiction, a conflict between the system of technique, which augments according to its own laws, and the society dominated by technique. To follow a comparison that I employed to shed light on the relationship between technique and the system of technique, it is almost like cancer developing in a live organism. But I do not mean to say that technique is a cancer; this is just an analogy to present the problem more effectively. Cancer, the cancerous cells, proliferate according to their own law. Cancer increases with its own specific dynamism; and it does so within a live organism, within a different set of cells, which obey different laws and which will be disturbed, sometimes completely unbalanced and disrupted, by the development of cancer.

The system of technique is rather similar in that it is located inside the society dominated by technique. Hence, one may say that wherever the system of technique advances, there is a greater disturbance of the social milieu and the human groups. In other words, there is a growth of what might be called a certain disorder, a certain chaos. Hence, contrary to what we might imagine, technique is quite rational, the system of technique is quite rational; but it does not subordinate everything to this rationality. There continue to be areas that are absolutely not subject to the system of technique; hence, some kind of crisis occurs. That is why I simply do not believe in the possibility of Huxley's brave new world. What we actually observe is a technical order, but *within* a growing chaos.

Will this state of affairs continue? Does this situation have no solution? As a matter of fact, we do not see any possible historical solution. It is quite simplistic, quite elementary to say that "we have only to adjust to technique" or "society has to be organized according to technical means." What this actually signifies is that during the five hundred thousand years of our existence, we have developed in a specific direction, and now we are suddenly being asked to change. Well, I am simply saying that we cannot suppress half a million years of evolution in a few short years. What we can predict for sure is that if the growth of technique continues, there will also be a growth of chaos. This does not at all mean a void or a crumbling of societies, but difficulties *will* increase.

Let us apply what we just learned about the milieu. We have developed (and here I might allude to Toynbee's theory of challenge) only when encountering challenges, only when meeting new circumstances to overcome. In a sense, the new challenge to us is our own invention, namely technique. But this is not necessarily negative. We are called upon to surmount technique just as we have surmounted the difficulties of society or the difficulties of primitive nature. In short, this is an expression of life, for life is a series of imbalances successfully restored to a state of equilibrium. Life is not something static that has been organized once and for all. Hence, this challenge of technique may be positive so long as we fully understand that it is a challenge to be overcome and that it is a fundamentally serious issue.

If, for instance, human beings had not taken seriously (and no intellectual interpretation was necessary) the challenge posed by cave bears, then they would very simply not have survived. Then, it was an immediate challenge, which they experienced constantly. At present we are obliged to travel a long intellectual road in order to understand the crux of the matter.

Given the current extent of relations between the system of technique and the society dominated by technique, and the extent of the true problems raised by technical development, no political action in the normal, strict sense of the term is adequate today. On the one hand, the politician and the political institutions are totally incapable of mastering technique. They are incapable of normalizing the techno-social phenomena and steering them. Our institutions were invented between the seventeenth and the eighteenth centuries, and they are adapted to situations that have nothing to do with what we now know. One need merely recall the total impotence of the legal system in fighting pollution. Obviously, we can always issue decrees and pass laws, which sufficed for the problems of society one hundred years ago. But none of this is effective against pollution, and I could multiply the examples along these lines.

Likewise, as we have already said, the politician is totally unfit for technical problems. But just as we cannot master technique, so too the politician cannot rationalize behaviour or find a new organization for society. For this would require the most totalitarian and the most technical government that has ever been imagined. However, we are not about to create such totalitarian or technical governments. At most, we have political authorities that are gradually and with difficulty adapting a few old governmental methods to new instruments. Indeed, when politicians realize the full scope of the problem, they become totally impotent. Hence, I believe that politicians can change neither technique nor human beings and society. In any case, for the challenge now facing us, we cannot expect any response along the road of traditional politics.

Politics is in no way acting upon technique and its problems. It is actually providing a framework for the events and trying to respond to the circumstances. In short, there is no such thing anymore as large-scale politics. It is quite astonishing to see the extent to which the great ideological systems—for instance the Communist systems in both the Soviet Union and China—have vanished, giving way to step-by-step policies. The USSR and China are totally falling in line in terms of the development of technique, and are therefore in the same situation as the Western world. Indeed, I believe that modern society has two entirely different, entirely distinct levels: the level of appearances and phenomena; and the level of structures.

In appearance, there are many movements, changes and events. Not so long ago the World Council of Churches investigated the question: "What is Christianity becoming in a changing society?" As if—and we all believe this—as if change were the fundamental trait of our society! The only thing that is really changing is appearances. It is obvious that the Soviet influence, say, in Africa, is tending to replace the Chinese influence of the nineteen sixties. Granted, this is not unimportant. But ultimately, the Chinese and the Soviets are more or less doing the same thing. Hence, we witness a large number of events which always boil down to a certain number of rather simple elements. The surface may seem very agitated, but the depth remains extremely stable. One can draw a well-known comparison to the ocean: the surface may be extraordinarily whipped up with waves and a tempest; but if one descends fifty metres, everything is calm, nothing is stirring.

Sociologically, I would say, we actually have three levels: the level of events and circumstances, which is always the level of politics; the level of far-reaching changes, for instance economic phenomena, which are longer-lasting and less circumstantial; and the level of stable structures, which, I believe, are given us by technique.

Technique fundamentally structures modern society. It is not that technique does not change. When I say it is stable, I am not saying it does not change. But it obeys its own law of evolution, and it is only very slightly influenced by events. It can be limited in its own development. Clearly, in the world we live in, we do not know everything that technique would allow. Blockages crop up—for instance, economic ones. In France, we know the contradictions in the National Health Service. The costs are so high that a choice must be made between extremely sophisticated medical techniques and an increase in hospital beds for the most common illnesses and operations; we cannot have both. Hence, in the basic structures, there are blockages coming from the two other levels.

However, there is no fundamental change. Technique does not obey events in any way. Yet obviously, what interests us, as people taken with information, with the news, with everything exciting and fascinating, is the events. However, the more fascinated we are by political circumstances, speeches, and ideologies, the more we leave the structures free to function as they do. We can focus on an important political discussion about the Third World, but in reality, the power of technique expands in regard to the Third World too—and this we do not see. We are so excited by events, by circumstances, by the latest news, that in regard to fundamentals, we always feel we have time. Even if we do not understand the stakes of the game in regard to technique, we always feel we have a great deal of time ahead of us. But this is not true. If technique keeps growing, then disorder will keep growing; and the more disorder increases, the greater our fundamental danger.

May one say that there is no help, no hope, that all is lost and we can only let things happen? By no means! I think that humanity—as I have already said—has frequently been challenged and endangered in an equally fundamental way, and at first sight, people saw no way out. In 1935, we saw no way out from the Hitlerian dictatorship. It was something terrifying, on which we seemed to have no grip. Likewise, those who were critical of Stalinism saw no way out. We were convinced, myself included, that things would go on in exactly the same way after Stalin's death. All the same, there were a certain number of changes. Hence, we may not see any way out for now, but we should not claim that none exists.

I feel that, in any case, there are groups who hold out some hope. On the one hand, the groups from certain milieus that express the chaos in the midst of which we live. That is, the groups, the milieus of a certain age—youths, for instance—who feel the shock of this society most strongly and most harshly and who tend to reject it, even if, for the moment, no solution can be found.

Then there are the groups who are beginning to be conscious of what is happening. I will limit myself to discussing the antinuclear movements, because all this is very well known. The technical validity or nonvalidity of their arguments does not matter. The important thing is to be capable of posing the problem on the most basic level. Even if one can affirm that the nuclear plants are totally harmless, the real question is one of society's choice; and the antinuclear groups would therefore be right. Likewise, the ecological movements, the consumer movements, the neighbourhood associations. The latter are citizens' groups who feel that we don't get rid of problems just by electing a local government. After all, the city council can only run the municipality. Thus, we have groups who feel that everything concerning their neighbourhood life is of interest to them, and they ask to receive all documents, they discuss all the decisions of the municipal council. They are capable of arousing public opinion in certain cases. Generally, they form a mechanism that I might call a spontaneous referendum. I find this a new phenomenon and a very important one in the political world.

Then, we have to take the women's movements into account. They strike me as extremely serious and fundamental—so long as their objective is not to become masculine! That is, so long as women understand their specific role and do not wish to play the same role as men in the same work, the same framework, and the same techniques. If women become men, what is gained? On the contrary, what strikes me as fundamental is that in a society in which the masculine extreme is crystallized in technique, the feminine part, which, I would say, is focused on sensitivity, spontaneity and intuition, is starting to rally again. In other words, I feel that women are now far more capable than men of restoring a meaning to the world we live in, of restoring goals for living and possibilities for surviving in this world dominated by technique. Hence, the women's movements strike me as extraordinarily positive.

In this list of groups, I have not mentioned the proletariat or the Third World. In European countries, thoroughly permeated with Marxist thought, the proletariat was the bearer of hope for the world because, even without precise knowledge of Marx, people saw the proletariat as the most wretched, the most "alienated," people who would be forced to revolt in order to wipe out their own inhuman condition. The proletariat is, in general, thoroughly integrated in the world dominated by technique by organizations like trade unions or political parties having purely industrial views and goals, and by situations that involve the proletarian in technique. Hence, the proletariat still thinks about issues in terms of the social and economic situation of the nineteenth or early twentieth centuries. Movements like trade unions do not see the new problems at all. For now, at least, and until a new consciousness is reached, I do not believe that the proletariat offers a future for humanity, any more than the Third World does.

We have already indicated that the Third World has progressively lost its specificity as the techniques introduced in those countries upset whatever was unique and singular about their cultures. I think that it is a mistake to investigate the transfer of technique. It is not enough, as is all too often said, to act with great care, to seek ways of adjustment. The transfer of technique can take place, and individuals and even certain groups in the Third World can be psychologically adapted. But in reality, the shock of technique causes a total breakup of the society. Hence, new studies on the transfer of technique will not solve this problem. The question is whether the civilizations of the Third World—India, Islam, and so on—being totally different from the Western world, are capable of absorbing Western techniques and integrating them into a totality of culture and civilization that is utterly new.

The shock of absorbing techniques has apparently destroyed the specific character of most of these societies. When one tries to redis-cover the cultural roots, they seem so backward and impossible that, in the eyes of all humanity, one is dealing with an absurdity. I am thinking of what has happened in Iran with the Ayatollah Khomeini—his desire to return to a pure, hard Islam, indeed to the Middle Ages, with a rejection of all techniques, which is unthinkable and unacceptable. There is no integration of techniques into a society with a different culture. It is an either/or situation: either technique or our Islamic society. That is the conflict of Iran today. Obviously the Ayatollah Khomeini's position is absolutely untenable. He is bound to be defeated because one can no longer live without accepting tech-niques. Iran will have to renounce the specific nature of an Islamic society.

There is, however, a further element which makes me feel that the Third World is no longer a resource in regard to the challenge facing us. You see, the very mentality of the inhabitants of the Third World has been transformed. On the one hand, the elite have only one idea: to develop technique, to enter the mainstream of technique. Both intellectuals and politicians are fascinated with this notion, just as the rich of the Third World are interested—in the most banal sense of the term—in developing Western techniques. In both cases the goal is to enter the circuit of Western technique.

On the other hand, for the poor in the Third World, technique clearly seems like a hope, the hope of overcoming poverty. In the mythology of the Third World, technique has succeeded in making the West rise from its own poverty. Therefore, they believe all they have to do is adopt Western techniques, and they too will profit from this development. One cannot contradict this notion, in the light of how poor and wretched the people of the Third World are. But they fail to realize that they are launching the twofold process of destroying their culture and entering into a universe that is totally alien to them, a universe that will bring disruptions on a psychological level and that will in fact cause, in all areas, far more serious disruptions than in the Western world. The West has adjusted gradually to its development of technique—and we know how badly and with how much difficulty. It has taken us two hundred years. How then can the Third World endure the shock, psychologically and sociologically, when it is asked to absorb this technical apparatus and this system of technique in just a few years?

Within this international framework, and especially considering what we have just said about the Third World and the gradual destruction of its unique cultures (despite the ideologies of, for example, Africanism), we must, I believe, realize that the true powers in our time are no longer the rich countries or the populous ones, but those possessing the techniques. The term "rich nation" instantly brings to mind the Arab countries with their oil. Of course, these countries do impress us greatly with their influence on all economic and political life. In fact, however, the accumulation of their wealth is not bringing any true interior development or any sort of independence from the West.

It is, I feel, very important to realize that these riches do not permit the emergence of a new type of society. They simply allow the adoption, the purchase of what the West has already done. One need only think of the very characteristic example of buying ready-made factories, delivered "key in hand," so to speak, and set up in the Arab countries. What is this? In fact, it is the implantation of Western techniques in the Arab world. Likewise, in the terrible war between Iraq and Iran, everything is Western, including the materials and the strategy. Nothing remains of Arab military culture.

Hence, the wealth of Arab countries does not give them real power. The countries with real power are those that have the technical instruments, that are capable of the technical progress that is confused with *development*. It is not real development, but simply growth, a growth of power. We ought to recall the difference that many sociologists and economists make between *growth* and *development*. Schematically, we may say that growth is chiefly quantitative and development qualitative. In an economy, aiming at growth means trying to produce more cement, more iron, or more wheat. Aiming at development means looking for the most balanced and least harmful economic structure, recognizing the value of the statement "small is beautiful," and achieving higher quality in consumption.

This distinction between growth and development obtains equally for politics and societal organization as well as for economics. So far, however, technique has always emphasized growth and the growth of power. And this power is both economic and political, of course.

By the same token, a large population does not imply real power. (This is the problem of the Third World.) People never stop emphasizing the dreadful injustice that exists because of the difference in standards of living between the Western world and the Third World. But this difference is accentuated by the very rapid advance of technique in the Western world. It is not simply the dynamics of capitalism, but rather the development of techniques. Hence, the axis of power is determined by the progression of techniques. At the same time, however, these techniques entail certain similarities. In order to exploit and to utilize techniques as much as possible and to maximize their yield, we must be able to organize society in a certain way, we must be able to put people to work in a certain way, and we must get them to consume in a certain way. Hence, the ideological oppositions are growing less and less important. The ideological and political conflicts in the strict sense of the term are rendered obsolete by the identical nature of the techniques.

Techniques are pretty much the same in the Soviet Union, the United States, and Europe, with only slightly different rates of growth. China is now moving in the same direction, evolving in the same manner, and is attempting to technicize progressively. As a result, political structures are growing more and more alike, as are economic structures. It is no coincidence that the Soviet world is beginning to talk about a market economy, a natural formation of prices through competition. Not that the capitalist system is better; rather, both sides are looking for the best forms, the most effective ways of using techniques. Likewise, the Western world is talking more and more about economic planning. Hence, an obvious convergence, with identical objectives, namely technical power, and the domination and utilization of raw materials for technique. Ideologies no longer count. Whether the discourse is Communist or capitalist, liberal or Socialist, in fact, everyone is obliged to do more or less the same thing.

I could give countless examples of these facts. For instance, when the Swedish Socialist Party was beaten by an antinuclear platform, the Liberal Party, on coming into power, realized it simply could not carry out the electoral promises it had made. Technique won out, and Sweden was forced to begin constructing nuclear plants.

This example shows the convergence of the technically powerful nations; however, this convergence does not automatically guarantee peace. All we can say is that ideological politics is now secondary, and that the conflict between the powers comes from an excess of power, an excess that extends beyond the national boundaries. In the past, people offered long explanations for the conflicts between capitalist nations, saying that capitalist production had to conquer new markets throughout the world. Hence, it was economic output that caused wars. Now, however, the risk is obviously the excessive power of the three (and soon four) great creators of technique. They will soon find themselves facing one another in such a way that a conflict will be inevitable—the conflict over the use of raw materials, for example. It is a question of life and death. This, ultimately, is what endangers world peace, and nothing else.

My interpretation of the phenomenon of technique as a milieu, as a system, has led me to get involved in society, as I tried to explain earlier. However, it was never my goal to go back, to declare that technique must be eliminated. I was looking for a new direction. So I tried to reach what is known in France as "the base" of society. "The base" is the average person, the one who simply lives his or her own life, who has no great ambitions or special intellectual development; but who still has something like spontaneity, openness, often allowing him or her to understand the things that are happening, so long as they are shown, and to understand them in such a way that he or she is relatively better prepared than intellectuals, technical experts, and executives to take the values of life seriously. All this led me to concentrate on local initiatives—that is, to rely on direct and close relationships to form groups for investigating the issues that require people to take a stand on technique and the system of technique, but which are also very concrete.

Let me give you an example of ecological action in the region of Aquitaine. I tried to get intellectuals to develop a critical attitude so that they would question the very techniques they were studying. These intellectuals included scientists, lawyers and administrators. The point was not to reject administrative technique or juridical technique, but to clearly know what we were doing by employing them; to know the visible, immediate results and the secondary and less visible drawbacks. In other words, very close attention must be paid to any technical interference in the social or psychological domain. It was a great consolation for me to see people not going backward but realizing that the most highly developed technical means are not necessarily the best, even though they are the most efficient. I am thinking of insights shown by the doctors I worked with. They saw that many tests, although highly developed from a technical point of view, are ultimately no more certain than the diagnoses that were once made by more elementary procedures, but demanded greater personal commitment from the physician. In other words, a very large number of laboratory tests and clinical examinations are absolutely useless. They are technically highly developed, but often very dangerous and sometimes very painful. Ultimately physicians and surgeons (I am speaking of the most highly qualified) recognize that the results and the knowledge attained are no greater. This is an example of a critical stance in regard to the very techniques we use.

At the same time, I was obliged to remain on the fringes with all my activities. Again and again, people tried to draw me into political circles, saying that something was happening politically that might lead to the acceptance of my analyses! This is a trap for the ecological movements. I feel that any action pertaining to the milieu of technique must remain on the fringes because this milieu is extraordinarily enveloping and, I might say, extraordinarily seductive. My work, therefore, is obviously on a small scale; it requires much effort for apparently meagre results. While crowds of people adopt all the technical developments, we can act only on individual levels. Hence, this is a true artisan's work. Nonetheless I am fully convinced that my slow labour, involving small numbers of people, is actually a point of departure for an internal change in society. To use big words, confronted with the phenomenon of technique and the new milieu we live in, we must have "mutants." Not the mutants of science fiction—the technical human being with a robot's brain—but quite the opposite. To be a mutant a person needs to become someone who can use techniques and at the same time not be used by, assimilated by, or subordinated to them. This implies a development of the intellect and a development of consciousness which can come about only for individuals, but it is the only development possible.

This leads, obviously, to the problem of educating children. For a longer or shorter period, our children and grandchildren, we must realize, will be living in a technical milieu, and we cannot even for one second imagine that we can raise them without some contact with it. Once again, the point is not to refuse to admit that technique exists, because it does exist; it is our milieu.

This goes back to what I was saying about the milieu. I know that it has in fact happened that when historical societies organized, small groups or sometimes individual people absolutely refused, saying: "We want to keep living like monkeys in the forest." Of course, they could do so, rejecting the development of society. But this was no solution. Those who continued living in the forest became extinct.

In the same way, one cannot claim that we can go on living as in the nineteenth century. We cannot bring up our children as though they were ignorant of technique, as though they had not been introduced from the first into a world dominated by technique. If we tried to do that, we would make total misfits of our children, and their lives would be impossible. They would then be highly vulnerable to the powers of technique. Yet we cannot wish them to be pure technical experts, making them so well fit for the society dominated by technique that they are totally devoid of what has until now been considered human.

Hence, I think that on the one hand we must teach them, prepare them to live *in* technique and at the same time *against* technique. We must teach them whatever is necessary to live in this world and, at the same time, to develop a critical awareness of the modern world. This is a very delicate balance, and we should not delude ourselves. We are preparing a world that will be even harder to live in for our children than it is for us. For us it is already complicated. And our children will be forced to deal with even more difficult situations.

Let me tell you of an experience that strikes me as dreadfully enlightening in its cynicism. I am rather well acquainted with the president of Electricité de France (the French national utilities company which is also responsible for the nuclear power plants). I was talking to him, discussing the dangers of nuclear plants point by point. Finally, in regard to two items in particular, he acknowledged that there were indeed some insoluble problems. And then he made the following extraordinary comment: "After all, we have to leave some problems for our children to solve."

That is the cynical attitude of the technical expert who knows his limits; it reveals that our children are indeed going to have difficult problems. Hence, in the immediate future, I feel that our children should be like all the others, go to the same schools as everyone else. But, at the same time, we should try to set up an alternative school, as it were, a parallel institution, where children learn to live differently and, on an existential level, learn to question the certitudes taught them in regular schools. Of course, this can be done only in communities of parents. One simply cannot provide such an orientation for life in a purely familial framework; and one cannot do work of this sort all alone with one's own children.

4
Faith or Religion?

THROUGH OUR ANALYSIS of the system of technique, we are obviously led to reflect on two kinds of issues: the human condition in this system; and the conditions that are necessary for taking the positions or making the critiques that I have presented.

Regarding the human condition in this system, we have repeatedly witnessed the transformation of human beings by technique. We must understand that, no matter what the political form, no matter how developed or underdeveloped a country may be, all the citizens agree on the development of technique, notwithstanding the dangers, and people offer justifications—ideological, intellectual, or philosophical. Let me quote a few passages from my book *The Technological Society.*

> It is literally impossible for the public to believe that so much effort and intelligence, so many dazzling results, produce only material effects. People simply cannot admit that a great dam produces nothing but electricity. The myth of the dam . . . springs from the fact that mass man worships his own massive works and cannot bring himself to attribute to them a merely material value. Moreover, since these works involve immense sacrifices, it is necessary to justify the sacrifices. In short, man

creates for himself a new religion of a rational and technical order to justify his work and to be justified in it.

* * *

Never before has so much been required of the human being. By chance, in the course of history some men have had to perform crushing labors or expose themselves to mortal peril. But those men were slaves or warriors. Never before has the human race as a whole had to exert such efforts in its daily labors as it does today as a result of its absorption into the monstrous technical mechanism—an undifferentiated but complex mechanism which makes it impossible to turn a wheel without the sustained, persevering, and intensive labor of millions of workers, whether in white collars or in blue. The tempo of man's work is not the traditional, ancestral tempo nor is its aim the handiwork which man produced with pride, the handiwork in which he contemplated and recognized himself.

I shall not talk about the difference between conditions of work today and in the past—how today's work is less fatiguing and of shorter duration, on the one hand, but, on the other, is an aimless, useless, and callous business, tied to a clock, an absurdity profoundly felt and resented by the worker whose labor no longer has anything in common with what was traditionally called *work.*

This is true today even for the peasantry. The important thing, however, is not that work is in a sense harsher than formerly, but that it calls for different qualities in man. It implies in him an absence, whereas previously it implied a presence. This absence is active, critical, efficient; it engages the whole man and supposes that he is subordinated to its necessity and created for its ends.

* * *

Consider the average man as he comes home from his job. Very likely he has spent the day in a completely hygienic environment, and everything has been done to balance his environment and lessen his fatigue. However, he has had to work without stopping and under constant pressure; nervous fatigue has replaced muscular fatigue. When he leaves his job, his joy in finishing his stint is mixed with dissatisfaction with work as fruitless as it is incomprehensible and as far from being really productive work as possible. At home he "finds himself" again. But what does he find? He finds a phantom. If he ever thinks, his reflections terrify him. Personal destiny is fulfilled only by death; but reflection tells him there has not been anything between his adolescent adventures and his death, no point at which he himself ever made a decision or initiated a change. Changes are the exclusive prerogative of organized technical society, which one day may have decked him out in khaki to defend it, and on another in stripes because he has sabotaged or betrayed it. There is no difference from one day to the next. Yet life is never serene, for newspapers and news reports beset him at the end of the day and force on him the image of an insecure world. If it is not hot or cold war, there are all sorts of accidents to drive home to him the precariousness of his life. Torn between this precariousness and the absolute, unalterable determinateness of work, he has no place, belongs nowhere. Whether something happens to him, or nothing happens, he is in neither case the author of his destiny.

The man of the technical society does not want to encounter his phantom. He resents being torn between the extremes of accident and technical absolutism. He dreads the knowledge that everything ends "six feet under." He could accept the six-feet-under of his life if, and only if, life had some meaning and he could choose, say, to die. But when nothing makes sense, when nothing is the result of free choice, the final six-feet-under is an abominable injustice.

* * *

For an hour or two, he can cease to be himself, as his personality dissolves and fades into the anonymous mass of spectators. The film makes him laugh, cry, wonder, and love. He goes to bed with the leading lady, kills the villain, and masters life's absurdities. In short, he becomes a hero. Life suddenly has meaning.

The theater presupposed an intellectual mechanism and left the spectator in some sense intact and capable of judgment. The motion picture by means of its "reality" integrates the spectator so completely that an uncommon spiritual force or psychological education is necessary to resist its pressures. In any case, people go to the movies to escape and consequently yield to its pressures. They find forgetfulness, and in forgetfulness the honied freedom they do not find in their work or at home. They live on the screen a life they will never live in fact.

It will be said that dreams and hope have been the traditional means of escape in times of famine and persecution. But today there is no hope, and the dream is no longer the personal act of an individual who freely chooses to flee some "reality" or other. It is a mass phenomenon of millions of men who desire to help themselves to a slice of life, freedom, and immortality. Separated from his essence, like a snail deprived of its shell, man is only a blob of plastic matter modeled after the moving images.

We must, I feel, simply focus on two decisive elements. On the one hand, we have emphasized that Western society, modern humanity, is faced with all the problems raised by technique. This is a challenge for the entire human species—a final challenge when we consider the risk of atomic war, and the most serious that humanity has encountered since the beginning of our history. Confronted with these problems and this challenge, modern humanity is powerless. We are in the rather dreadful situation of realizing that the danger is extreme, the problem complex, and that we have no way to deal with it, nor do we know of anyone who can find a way. We have no means because we were not trained for this and because those in charge—we have spoken of the politicians—are obviously left behind by what is happening. Hence, a first contradiction in our situation.

The second contradiction, which strikes me as highly characteristic of Western society, is that we are all subjected to disciplines that are getting more and more rigorous, more and more severe. But these are external disciplines. As government administration, for instance, keeps improving, order and discipline have to become stricter and stricter. To point out something very simple: disorder is more and more unacceptable in the street if traffic is to keep moving properly. Hence, we are subject to strict discipline in a society that, at the same time, has lost its values. That is, we feel less and less that this discipline is indispensable. The harder, stricter, sterner this discipline becomes, the less we recognize that it has a reason for being, because in order to regard a social discipline or control as indispensable, we must acknowledge its value. This society, however, has lost its values, has lost its meaning. Hence, every one of us is always ready to challenge anything, to reject all the disciplines. At the same time, the more technique advances, the more extreme the disciplines we are subjected to.

These two contradictions, which I feel are the most basic of our time, cause something that is observed constantly: the situations of anxiety and neurosis, which characterize many in our society. We need to realize that we are more anxiety-ridden than ever before in history, that neuroses are more numerous—most likely not because we in the modern world go mad more easily, but because we are confronted with particularly arduous, particularly difficult situations. Hence, a first set of questions faces us. Is the human condition hopeless? Is there some way of coping with it? What can be said and what can be done?

We have also noted that this analysis of the system of technique led us to reflect on the conditions required for the critical stance that I spoke of previously. A critique requires an outside reference point; one can criticize something only in accordance with a scale of values. If one has no point of comparison, no scale of values, then obviously one can judge nothing. Let me point out in passing that when I speak of "critique" like that, it must not be understood only in the negative sense. In keeping with a term now often employed, this is "constructive criticism," in other words, it is a matter of distinguishing, of discriminating. This is the etymological meaning of "critique": to discriminate between what can be retained and what must be rejected.

This can be done only if we have a stable given, which allows us to form a judgment on the situation. What would our stable givens be? "Man"? Very often, he is indeed the subject when one reads: "'Man' does this" or "'Man' could judge that." And I myself admit that I have used this facile expression on occasion. Actually, however, we know that man is a fleeting reality. It is difficult to maintain that a permanent human nature exists. We see so many variations in different societies and so many variations in history. It is not human beings as such who allow us to use them as the critical point of reference, and the same holds true for *history*.

History allows any interpretation whatsoever. It is one of the faults or one of the dangers of our time to submit everything to history, to regard nothing as fixed. History, having become the ultimate phenomenon, eliminates the basis for a criticism and the points of comparison.

Finally, a further reference point that is often advanced: Marxism. But I cite it only as a reminder, since I have already emphasized that Marxism has been totally integrated.

Hence, we need a reference point. But which? We also need a viewpoint. We have to locate ourselves on the outside in order to look at the phenomenon. If, for instance, I want to know the speed of a train and I am inside the train, with no exterior viewpoint, I can know nothing for sure. I have to have a viewpoint outside the train so that I can watch the train pass; or else I have to see some fixed, stable outside point from within the train, allowing me to evaluate the speed. Where can we situate ourselves in order to have a viewpoint for looking at technique? Where can we situate ourselves outside of technique, if technique is the organization we have spoken of, with its tendency to be all-encompassing and totalitarian? We are, it seems, so deeply incorporated in technique that we absolutely have to find a different place from which to look at it.

I might then allude to that frequent habit of intellectuals today, the reference to madmen, neurotics, paranoiacs, schizophrenics which has become the exterior viewpoint for situating ourselves in order to look at what we are and what the world is in which we live. We must realize that a choice of this nature is a truly desperate one. To say that ultimately a reference to paranoia or schizophrenia allows us to understand what is happening in our society is more or less to declare that we have become totally incapable.

Not only must an awareness of what is happening in our society be possible, but it must also be bearable and tolerable. It has to be bearable even if it reveals things that are extremely harsh, extremely severe. And it is true (an objection that I have often encountered) that to describe the system of technique as I have done may lead to an awareness of a certain reality, but one which is so menacing as to leave us discouraged and despairing. In other words, if technique is indeed as I have described it, then there is nothing we can do. We merely throw up our hands and, ultimately, we can only commit suicide.

In relation to two issues, namely the human condition and the conditions necessary for a critical awareness, it is Christian faith and the Revelation that intervene for me. Here we come to the point of dialogue between the two parts of my book: the sociological part and the part on Christian reflection or theology. These two elements are closely linked, because on the one hand it is only by living this faith in Jesus Christ that I could do this analysis of society, and on the other hand, my analysis of the world dominated by technique demanded a more and more vigorous faith from me and an increasingly exact theological knowledge.

Before we tackle the problem of Christian faith and Revelation in regard to the issues we have already indicated, some light must be shed on the matter of religion.

Religion is a natural phenomenon, a spontaneous phenomenon. For a long time, it was said that humans are religious animals. This statement should not be taken as a defence of religion or Christianity. Historically, and in our societies more than elsewhere, people have tried to destroy religion. The rationalist nineteenth century, for instance, in Europe and especially in France, claimed to do so by envisaging a purely scientific and purely rational human being. But when people make such efforts to destroy religion, it simply reappears elsewhere. In a world where the old, traditional religions— Christianity, Buddhism, Islam—are supposedly on the wane, we can observe the resurgence of religions like the so-called secular religions, characterized by political types of faith. We know that Hitlerism was a religious political phenomenon. Stalinism too was a religious phenomenon, and so was Maoism. (Chinese factories had small altars, genuine altars with a portrait of Mao, candles, and incense. Workers began the day with a sort of religious salutation, which was a kind of prayer and worship for Mao, with utterly classical religious forms.)

Furthermore, in our so-called rational and laicized societies, we are witnessing a return to primitive faiths. In France, for example, there is a proliferation of astrology, fortune-telling, and horoscopes.

There is also a whole set of beliefs in extraterrestrial beings. It is quite amazing to watch people who call themselves totally rational and even totally scientific and yet are so nervous or worried about the possible presence of extraterrestrials. These phenomena are totally religious. In addition, we must mention things like drugs or popular music, which are likewise typically religious.

In other words, we apparently cannot escape human religious expression. In a society, religion has extremely well-known, extremely precise functions. It serves to hold a society together. When one destroys a religion, one will see the social group come apart. Religion gives people a kind of overall explanation of the world, which is very important: one cannot be satisfied with a purely logical and rational science, a science that knows it is limited. This is why the scientist acknowledges that science is limited. Although ordinary people believe in science on a religious level—what used to be known in France as "scientism"—science is unable to give us an overall explanation. We need to know where we come from, where we are going, how we are situated, what our future is. These answers are furnished by religion, and religion alone; otherwise, humanity is completely lost.

Religion also serves to encourage us to live. It is not easy to live; we need to be encouraged and helped. All this is purely sociological, purely psychological, purely natural. Religion on this level has no sort of necessary reference to a God—a truly transcendent God. On the contrary, the God in question has to be very close by. That is why we have all possible physical representations of our divinities, so that they can be seen. In other words, the existence of the religious feeling does not mean that God exists.

I would therefore like to establish a difference and even an opposition between religion such as I have just described and the Christian Revelation. I prefer to call it Christian Revelation or Christian faith rather than Christianity, because the suffix *ity* already implies a shift to the sociological. Following Karl Barth, I would like to show the difference between the two. The Revelation given to us in the Bible, the Revelation that is in Jesus Christ, refers us to a God who is different from all other gods, a God who was called the "Wholly Other." He is totally different from us, and we can understand nothing of him with our human means, our intelligence, our scientific means, or our feelings. We cannot know anything of Him, except what He reveals to us of Himself. We must understand that if God is truly God, then He is alien to us, He is different, He is not within our reach, and we can know of Him only what He Himself tells us. According to a saying that was current in France at the time of Barth's greatest influence, "Only God can speak about God."

Hence, there is no common measure between this God and humanity, no common denominator but Jesus Christ.

Thus, the God revealed in the Bible does not in any way correspond to our spontaneous religious feeling. On this point, Biblical texts are very explicit, when, for example, the prophet speaks on God's behalf: "My thoughts are not your thoughts, my actions are not your actions." In other words, in its truth, the Revelation of God in Jesus Christ is truly the *opposite of religion*. This opposition between the Revelation of God in Jesus Christ and religion per se strikes me as bearing on two chief points.

First of all, religion comes from human feeling—an anxiety, a lack. Here we encounter all the explanations, for instance, of Feuerbach or Marx on religious feeling, which I believe are correct.

Secondly, religion tends to *rise* toward God. It is always the same pattern: the human soul ascends toward God and makes an effort to enter heaven in order to join God where He is. Hence, the countless religious monuments, all the towers, spires, cathedrals, the Chaldean ziggurats, which always rise toward God. This is one of the elements characterizing the Tower of Babel, and this is why it is condemned.

Thus we have two aspects of religion: it comes from a person's feeling; and it tends to portray a movement toward God. The Revelation of God in Jesus Christ, however, is the opposite of these two tendencies. It contradicts our religious feeling, and does not satisfy a person at all on this level. Why? When we examine the diverse religions, we observe that the gods invented by men or women are always gods in the service of them. People actually want to be the master of these divinities; they have to render services that are expected. We know that this leads to sanctions in the pagan religions as well as in certain aspects of the Christian religion; the god is punished when it does not respond according to people's wishes. In certain rural parts of France there existed a completely pagan custom of punishing a saint for not answering a prayer by turning the statue's face to the wall for being so wicked.

This is quite typical of the attitude: we wish to *use* the divinities. In the Bible, however, we find a God who escapes us totally, whom we absolutely cannot influence, or dominate, much less punish; a God who reveals Himself when He wants to reveal Himself, a God who is very often in a place where He is not expected and only rarely in a place where He is expected, a God who is truly beyond our grasp. Thus, the human religious feeling is not at all satisfied by this situation.

The other contradiction is that the Revelation that is in Jesus Christ and which manifests itself in the Incarnation, proceeds from above to below. Previously, we said that religion seeks to go from below, where we are, to above, where God is. But the Bible shows us the opposite. I am thinking of that great passage in the Epistle to the Philippians, which tells us what the Incarnation is: Christ is existing in the form of God and stripping himself of his divinity in order to become human; as a person, he takes the lowest position, that of a servant; and as a servant, he accepts the harshest and most humiliating punishment, that of slaves—the cross.

In other words, the road descends. God descends to humanity and joins us where we are. This is the opposite of the religious movement, in which people would like to ascend to where God is. Hence, we see a radical contradiction between all religions (I obviously can't go into minute details) and the fundamental path of Revelation.

However, Christians do not remain with the Revelation; they do not adhere strictly and exactly to their faith. Like all people, Christians also have religious feelings. Throughout history and the history of the Church and the history of Christianity, we have usually observed that those calling themselves Christian have always tended to transform the Christian *Revelation* into a Christian *religion*. That is what "Christianity" is. In it we find again the religious feelings, the rituals, the myths, with exactly the same structures as in all religions. Believers try to make Christianity take over the sociological functions of all religions. As a result, Christianity was said to be a religion like any other or, conversely, some Christians tried to show that it was a better religion than the others, although this is simply not the problem.

Hence, out of habit, because we live in a society, because we have religious feelings, the Word, revealed by God, was once again transformed into a religion, and people thereby attempted to take possession of God. The appropriation of God was, for example, attempted by the movement based on the theology of good works: designed to oblige God to save you because you are good. Another attempt was to place a value on sacrifices: I make sacrifices, therefore God must save me. We should not forget that a good number of theological undertakings were attempts to take possession of God. Theology claimed to explain everything, including the being of God. This is also the religious tendency: to try and explain God. It is of course possible to explain the Revelation that God gives us, so long as we know that beyond this Revelation, there is all the mystery of God into which we cannot penetrate.

People tend even more to transform Christianity into a religion because the Christian faith obviously places people in an extremely uncomfortable position—that of freedom guided only by love, and all in the context of God's radical demand that we be holy. "Be holy because I am holy," the Bible tells us. To avoid any misunderstanding, I should explain that holiness has absolutely nothing to do with the traditional Catholic notion of saints. Holiness means separation: I am a God separated from the world and from the other divinities; and you too are separated from the normal, habitual course of society and history.

This demand is extremely rigorous. Confronted with it, people try to transform this demand of freedom, love, and holiness into a morality. Hence, the Christian Revelation is transformed into religion, and the demand of God for our lives is transformed into a morality. Yet I would say that Christianity is ultimately an antimorality. We will see presently that this does not exclude the existence of a Christian ethics, which is not quite the same thing. Morality is a kind of catalogue of rules that one must obey. An ethics is an orientation toward life that calls upon us to develop all our possibilities.

As for the opposition between religion and Christian faith, one can likewise point out that this distinction can be made on other than theological grounds. We find a very interesting suggestion in Karl Marx. We know Marx's position on Christianity, the Christian Church, and God—*God* in quotation marks. Marx feels that Christianity has a purely sociological function in the superstructure of society. He holds that religion actually serves the ruling class to maintain its domination.

Nevertheless, toward the end of his life, in a letter to his friend Max Rugge, Karl Marx says something utterly remarkable: "When all the political foundations of religion are wiped out, when the organization and the institutional structure of the church are destroyed, then normally religious faith, the Christian faith would have to disappear. But it is not out of the question that the Christian faith will survive anyhow. This would mean that there is a religious reality, that does not depend solely on the sociological and the institutional; and under these conditions, we would have to heed this reality, which is not in the category of traditional religion."

This leaves us with a question. Marx says that after all, faith may be a fact, but we will not know it is a fact until the sociological trappings are destroyed; and if it is a fact, it must be heeded, for a good Marxist pays attention to facts. Thus we see how an honest intellectual position in Marx leads him to at least pose the question.

The deformation of Revelation and faith into "Christianity" has always occurred. At the moment, however, it is more serious than ever because of technique. There is thus a complementary impact of technique on the Christian phenomenon. This transformation can be analyzed along three lines.

First, it is certain that technique reduces Christianity to the inner life, to spirituality, to the salvation of the soul. This was already a widespread trend among Christians, but it is now heavily emphasized. Essentially, the argument goes more or less as follows: we technical experts do important things, on which depend the life of the society and of all people. We develop knowledge, methods, means, and power; everything else is superfluous and not very important. Naturally, we don't mind people having religious feelings, if they need them, as long as they don't interfere with the operating of the techniques—as long as they don't interfere, because the state is technicizing itself and becoming more and more rigorous in its demands. So these feelings must remain in the realm of the spiritual. This argument, incidentally, is extremely pernicious from another point of view. I have very often met technical experts who justify themselves as follows: "Thanks to technique, humanity will be relieved of all material and mechanical burdens, and it will be liberated for the spiritual life. So rejoice, Christians. We will take care of your basic needs, and you can rise to higher spheres." In fact this is a negation of the

Revelation, for if we take the Incarnation seriously, we cannot accept that the Christian faith is to be relegated to heaven. Jesus Christ came down to the earth. This implies an incarnation, and an influence on the concrete conditions of life.

A second aspect of this distortion of the Revelation by technique is a penetration of Christianity by techniques. I have two very different examples in mind. One is the techniques that are used to spread Christianity—for instance the techniques of propaganda or advertising. I do not, of course, wish to accuse anyone. But we can recall evangelical campaigns like Billy Graham's of several years ago. Organizations like this are purely technical; they are modelled on the great political organizations, and their aim is to spread Christianity, just as Stalinism, Nazism, and other movements were spread. It is exactly the same thing. We have to realize, however, that since the Revelation of God in Jesus Christ took a certain orientation, a certain form, it cannot be spread by just any method. There is a need to discern and evaluate the means, even when the technical methods are legitimized in advance.

The other example of this penetration of Christianity by technique is altogether different, namely structural linguistics used to analyze the Bible. This is a way of grasping the Biblical text with an extremely rigorous technique of reading, very different from older ways of exegeting. This technique is rigorously neutral and treats the text as an object without meaning, an object consisting merely of structures. We find a certain number of exegetes and theologians who are excited by structural linguistics. But in reality, they are killing the Biblical meaning, because the primary effect of this technique, like all other techniques, is to eliminate meaning altogether. Now if the Bible has no meaning, then why bother reading it? This impact is extremely deplorable.

Finally, there is a third aspect to focus on. People today frequently express a boundless admiration for techniques, a hope that we pin on their development. It's no use denying it. We are all convinced, for example, that thanks to improved medical techniques, cancer will soon be defeated. In other words, our hope is placed on an improvement of these techniques. We therefore now have an attitude in regard to techniques that is religious, whether in the area of absolutely unconditional admiration for the great works of technique or, as I have just said, in the hope placed in technique's development. Consequently, we have a kind of redirection of faith toward something other than the Revelation. I will give an example of this deliberate redirection, although this is obviously not happening today in the West. In the early 1920s, the Soviet government pursued the following antireligious propaganda in the schools. The teachers set up two flower beds. In one, they planted seeds and devoted all the necessary care to make them grow. In the other, they put nothing. But they prayed to God to make something grow and they told the children: "You see? When technique gives us the means to make something grow, then it grows. But when you pray for things to grow, nothing happens." Here we clearly have a redirection of faith toward something that is not the Revelation!

However, if we have succeeded in showing that the Christian Revelation is indeed the opposite of religions, this implies (since, as we have said, it cannot be set aside from what is happening in our time) that it has a role to play in the world. But this role is different from the sociological role of religions. This is the crucial point. When we transform Christian faith into a religion, when we turn it into "Christianity," then it plays the same role as any other religion. But if the Christian Revelation and faith are its opposite, then they have a different role according to the perspective of this Revelation. Let me present three major orientations of this different role.

First of all, the Christian Revelation offers both the criterion and the critical vantage point that we spoke of at the beginning, namely the reference point that is not incorporated into the system. If technique is total, if it is all-encompassing (that is, if the system of technique integrates into itself every phenomenon that arises), if it is "assimilative" (in the sense that all revolutionary movements are ultimately assimilated), then what can escape the system of technique? From a human outlook, we see nothing that does. We therefore need a transcendence in order to escape it. Only something that belongs to neither our history nor our world can do this. I mean, of course, something that does not "essentially" belong to them, because even the most distant planets are increasingly becoming part of our system.

We need a transcendence. When I say this, I am not being apologetic, I am not seeking to defend Christianity; that doesn't matter to me in the least. Nor does this need prove the existence of God. I simply mean to say that only one of two things is possible.

One possibility is that technique becomes our destiny, a kind of growing fate that grows and takes over all human realities. No culture will escape, as we have seen in the Third World. And, parenthetically, when I said that technique could trigger crises, I meant that there are determinisms during crisis just as there are determinisms during a time of equilibrium. Thus, the first possibility is that technique can become a true fate in the old and religious sense of the term by introducing an absolute determinism into our society and our world.

The other possibility is that something exists that technique cannot assimilate, something it will not be able to eliminate. But this can only be something transcendent, something that is absolutely not included in our world.

For the moment, I do not choose. I am simply saying that we are faced with this either/or. We are faced with either technique as our fate or the existence of a transcendent. The existence of this transcendent permits us to evaluate the world in which we find ourselves.

If this transcendent really exists, then let us suppose it is the one that was Biblically revealed in Jesus Christ. If He came down to us, then He is not included in our system. We can then place ourselves where He situates us, that is, in His transcendence. This then gives us the outside vantage point that permits the critique of the system. This also guarantees freedom, because there is no kind of freedom that we can claim to have in relation to technique. We need a freedom that is given to us from the outside. We need a freedom that comes not from us, nor from what we do. Only the transcendent in the system of technique guarantees freedom to humanity and a possible way out for society.

In other words, we completely reverse the traditional way of thinking about God. When we think about God in a very banal and ordinary way, we argue that if God exists and is all-powerful, there is nothing we can do. God foresees everything, hence, we cannot change anything. Our future is written. Indeed, the expression "it is written," which is common in Islam, has become widespread in Christian circles. It is written (or predestined). There is nothing we can do. This turns God into a kind of fate, which, I feel, is the worst misconception one can have of the Biblical God, because the Biblical

God is, above all, a God who liberates. He is not first and foremost a God who orders, commands, and constrains.

I will give only one example. The first Revelation that this God brings to Israel, in relation to which the Jewish people have reinterpreted their entire past, is the Liberation from Egypt. For Israel, God is first and foremost He who freed them when they were slaves and led them to a land of freedom.

I should point out something else, which shows to what degree these texts are existential in character. The word that is normally translated as *Egypt* in Exodus is *mitsraim*. But actually, this word designates more than just a geographical place. *Mitsraim* means: "the country of twofold anguish." The narrative therefore speaks about the real and political liberation of a people from bondage—and at the same time, about the liberation of humanity from the double anguish of living and dying.

This, then, is the first role played by the Revelation. God is He who liberates humanity and who liberates His people. This is confirmed with Jesus Christ, who is also the Liberator. When He speaks to us about the Holy Spirit, he speaks of it as the Liberator. When Saint Paul speaks to us about Jesus Christ, he tells us that Christ set us free. Thus, we meet this transcendent, *whose sole action is an action of liberating us, a liberation which is always begun anew. This liberation can be guaranteed and certain only if God is this transcendent.* Otherwise, He too would be encompassed in our system of technique. This is one role that strikes me as important for the Christian Revelation assumed in the Faith. It gives us the criterion, the critical vantage point, on the basis of this transcendence.

A second role, which corresponds to the issue we raised earlier, is that Christian faith gives us the possibility of viewing reality as it is without despairing, no matter how hard it may be. I have talked a great deal about one particularly threatening aspect of this reality as it is, namely technique. In giving this description of the system of technique, was I being pessimistic? I know that some critics regard my interpretation of technique as coming from Calvinism, that harsh, demanding, pessimistic religion with a predominance of sin! People are sinners, complete sinners; hence, whatever they do is bad; hence, technique is bad because it is made by them.

In response I would say that neither Calvinism nor the idea of sin has had the least influence in my investigation. In regard to Calvin, I have already said that I had long since rejected his ideas. I am in total disagreement with the Calvinistic way of thinking—double predestination, for example. As for sin, well, the further I advanced in my sociological and theological reflections, the less important the category of sin became for me. At twenty-two or twenty-three, before undertaking these studies, I was more influenced by the notion of sin than later on.

The basic notion of sin, as found in some preaching and in Calvinism, is that it encompasses everything, and that only when one has the terrible conviction that one is a sinner, does one learn the startling news that one can also be saved. I believe, however, that the Biblical Revelation is exactly the opposite. Once again, I have to credit Karl Barth with having seen that what the Bible announces is not sin, but salvation. It is only when people learn they are loved, forgiven, and saved—it is only then that they learn they were sinners. In other words, we can take sin seriously only by looking at Jesus Christ on the cross, because it is there that we learn the significance of sin. But it is by learning I am saved that I learn the importance of my sin. Consequently, this too is a message of liberation and absolutely not a message of gloom and condemnation for the human race. I believe that the theological development in all my later works, which try to show the universal salvation of humanity, demonstrates that the idea of sin had no influence whatsoever on my analysis of technique.

So, am I a pessimist? Not at all. I am not pessimistic because I am convinced that the history of the human race, no matter how tragic, will ultimately lead to the Kingdom of God. I am convinced that all the works of humankind will be reintegrated in the work of God, and that each one of us, no matter how sinful, will ultimately be saved. In other words, the situation may be historically dreadful; but it is never desperate on any level. Consequently, I can take the reality we live in very seriously, but see it in relation to salvation and God's love, which leaves no room for pessimism.

Those who object to this analysis of technique and who call it pessimistic are refusing to see the reality and are totally deluding themselves. I consider them bad doctors who do not tell a sick man what is wrong with him. Is a physician being pessimistic when he tells a patient that he has such and such an illness? The doctor has made the diagnosis, and he then bases his treatment on it. Would it be better for the physician to say: "It's not serious. Don't worry. Don't think about it. Just go on living as you have been," when the illness is serious? I feel that those people who would call the former doctor a pessimist are dangerous guides in a social reality such as ours.

Seeing this reality as it is, the technique that I have tried to depict, could truly be paralyzing and discouraging, and could lead to despair. But it is precisely here that the Revelation, accepted in faith, can bring promise, hope, and liberation. It brings promise in the sense that no matter how mad history may appear to us, it is situated within God's promise and it does lead to the Kingdom of God. It brings hope in the sense that this certainty permits us to live here and now. That is the answer to the anguish of modern humanity, and not only the anguish aroused by my analysis of technique. The anguish of modern humanity is subconscious, but it is fundamental. This is where we find the liberation from the double anguish, which I spoke of earlier. It is an answer to the anguish and the meaninglessness of living through a history that has no meaning. If history does have meaning, then this meaning is given to us by the Revelation and by Jesus Christ. When I say that this history has a meaning, one must not interpret this as saying that one can find a significance in every event and that one has to find what each one means. That is not what I am saying. Our history is our responsibility. But it is not "a tale told by an idiot." We are dealing with something that we know has a beginning and an end. When Jesus Christ tells us "I am Alpha and Omega, the beginning and the end," we know that history, with all its events, unrolls between two fixed points. But it is our task to make this history—and to make it with the courage given by the hope and liberation of Christian Revelation.

Finally, the third role of Christian Revelation received in faith is to remind us that beginning with the end, with this general orientation, which is the Kingdom of God, we must discern and evaluate not only our conduct but also our means. As far as our conduct is concerned, this is the traditional moral attitude. One judges a person's conduct or one's own by a certain set of criteria. But we must be careful here. While morality sets up *advance criteria* by which to judge, the Christian Revelation begins with the end. That is, one starts not with the Creation but with the *New Creation*; the Kingdom of God, with what awaits us at the end. This end toward which we go, which is revealed to us, allows us to return to where we are; and there we are able to judge not so much people's conduct, but the means. We now find ourselves at the very heart of the problem of technique, because the central issue today is that of the means and their power. We are constantly called upon in political life, in economic life, in our practical, professional lives, to voice opinions *on the means employed*, in other words, on all techniques. In order to be able to do this we must completely reject the frequently cited dictum that the end justifies the means. Historically and theologically, no end can justify the means. For—and this may be hard to understand theologically—the Kingdom of God, which is the end, the termination, the conclusion, is also a reality that is present now. The Kingdom of God is already present, here, in the means we employ. Hence, we must know whether or not these means are bearers of the presence of the Kingdom of God. Are these means of justifying humanity, means of liberating humanity, means of bringing joy to humanity, or not? Such is the judgment that we must make, knowing that no historic end justifies the means. We are gradually coming to realize this. The finest, the most grandiose, ends set up by political systems always translate into catastrophes. We also need to reject one of the characteristics of technique that we noted earlier, namely, the unlimited growth of the means without our being able to guide them toward human ends. It is precisely the Christian faith that leads us to reject this unlimited character of the means.

In sketching these three of many possible roles of Christian Revelation accepted in faith in this world, we need to clearly understand that this Christian faith implies on one hand that one acts in and does not simply try to escape the world, and on the other hand that everything is based on the fact that we believe.

But what do we believe? Certainly not in a Providence organizing everything for us. Certainly not in a God who does everything. And certainly not in the opposite: a God who acts by the agency of people and history. We must be quite clear that what we believe is that God's promise, received in faith, borne by us, truly changes the conditions in which we live and act. In other words, the presence of faith in Jesus Christ alters reality. We also believe that hope is in no way an escape into the future, but that it is an active force, now, and that love leads us to a deeper understanding of reality. Love is probably the most realistic possible understanding of our existence. It is not an illusion. On the contrary, it is reality itself.

It then follows that Christians ought to play a decisive role in the society dominated by technique, and I could simply refer to what the Gospels say: the salt of the earth, the leaven, and so on. In Christian circles, I generally note an authentic personal faith, which I am not challenging in any way. But in regard to the world, I perceive two situations: either a total indifference toward, and total ignorance of what is happening in society, and of everything for which we are responsible; or else, totally superficial involvement; for example, in politics.

Unfortunately, Christians do not seem to have any sort of lucidity when it comes to acting in society. In France, for instance, it's very fine, very virtuous, very *sympathique* to be left-wing, to be revolutionary. It shows a wealth of good feelings. But it is no different from being moralistic in the nineteenth century. The two are really identical: they show a basic lack of understanding of the world we live in.

In a society dominated by technique, what would it really mean for Christians to be bearers of the Revelation in faith? Here, too, I believe, one can focus on three aspects of the role Christians need to play.

First of all, one certainly should not reject technique. Mine is not an antitechnicism or a judgment against technique. It is not up to us to judge, because God alone is the judge. I have already said that the works of human beings are used by God to build the New Jerusalem. The book of Revelations says that the glory of nations will enter the New Jerusalem. The glory of nations also includes technique. Hence, our attitude is not antitechnical; rather, it is a critical acceptance of technique. I believe that this aspect can be seen in two domains.

For instance, we need to subject science and technique to the critique of the Revelation. We have now generally accepted that in the Revelation, the Biblical text, the sociological and psychological elements must be scientifically critiqued. We quite willingly accept scientific critiques of the Biblical text, but we should not forget the converse: scientific givens are never as certain as we imagine, and they too must be critiqued from a different point of view, from the standpoint of Revelation. Above all, however, our attitude will be what may be called iconoclastic. I do not make iconoclasm the principal and central characteristic of the Christian action, yet it is nevertheless important. Iconoclasm means the destruction of religious images, but what does it mean here? It simply means that we must destroy the deified religious character of technique.

If we see technique as nothing but objects that can be useful (and we need to check whether they are indeed useful); and if we stop believing in technique for its own sake or that of society; and if we stop fearing technique, and treat it as one thing among many others, then we destroy the basis for the power technique has over humanity. This, of course, is very difficult for technical experts to accept. If you check the ads in the press, you will see that they deify the world dominated by technique.

It is very difficult for technical experts to accept being the mere servants of machines that are relatively useful, but not all that important. I think, however, that this critical acceptance, combined with a necessary iconoclasm, is the first aspect of the role that Christians should undertake.

Secondly, Christians should be the bearers of Hope (*espérance*) in a society like ours—hope for people plunged in anguish and plagued by neurosis, which we have talked about; and hope for our history, for the future. But we must be on our guard. Christian Hope does not, as is too often said, consist in believing in humanity. It is precisely the contrary. Christian Hope means being convinced that we will not go along completely on our own. It is an affirmation of the love of God. In addition—a crucial element—Christian Hope, the Hope so fundamental to the Biblical texts, has a reason for being, a place, only where there is no more human hope. Human hope (*espoir*) is the feeling that tomorrow will be better. One may be in the throes of an economic crisis today, but one may have grounds for hoping that the crisis will be over in one or two years. So long as human hope of this sort exists, there is no reason for Hope (*espérance*). Human hopes will do. Hope, precisely, has no *raison d'être* unless there is no more reason for human hope. This is Hope against hope.

Hope will then simply be the fact that because God is God, because God is love, there is always a future. Even if today the future appears totally blocked, even if we no longer understand, even if we cannot foresee anything—which is certainly our situation—the future is possible and positive. It will not be catastrophic. In other words, bearing Hope means giving us courage to live today.

The third role of Christians consists in their being called upon as bearers of freedom in this society; bearers of freedom when technical conditioning is getting more and more rigorous, more and more determining for people. When we are told that we are set free by Christ, we must take it seriously. This means that no fate exists. It means that we must act like free people—in regard to what conditions us. Acting like free people means nothing in a free setting. But, since our situation is one of constraints and determinisms, it really means something to be set free by Christ.

Hence, there is a Christian ethics. I have already said that the Christian faith is the opposite of a morality, but that there is, nevertheless, a Christian ethics. But this Christian ethics is one of freedom; it is not theoretical or abstract. This leads to the obligation for a Christian to use every initiative to restore the possibility of people making their own decisions. These decisions are not necessarily on the religious level—whether a person converts or does not convert.

Already in our society, it is fundamental that people again be able to make their own decisions and be bearers of freedom. When things are thus conditioned, structured, and planned, there will always remain what we might call lacunae. There is free play between the parts of the mechanism, in these structures. The Christian is one who brings as much free play as possible into the parts in society—the government, bureaucracy, and so on—that are linked to one another. Flexibility is necessary. Also in politics it is very important—and this is not just a slogan—to work for human freedom. But we should not deceive ourselves, not convince ourselves that a movement is truly one of liberation when it is obvious that, once in power, it will oppress the people as much as the preceding government. Hence, we should, of course, take part in movements for human freedom but only if we are lucid and clear-sighted.

By no means do I think that these three orientations of Christian Revelation sum up everything that Christians may be called upon to do. Nor am I saying that the Revelation leads only to these three roles. I do feel, however, that these are roles we have to play and this is what as Christians we must live out in a society dominated by technique.

Appendix 1
Avoiding Some Common Misunderstandings About Technique

BY ADDING THESE TWO APPENDICES[1] designed to help you read Ellul, I am not in the least implying that his writings are unclear. Some of them were included in French high-school readers as examples of clear contemporary French writing. What I am suggesting is that the reading of Ellul's many works is difficult, for two reasons. First, there are many problems with the translations. For example, when I checked the translation of the interviews comprising the four chapters of this book, I felt compelled to make so many changes that my editor at CBC Enterprises simply could not believe me. We spent the better part of a day going over a sample of my corrections, and I had no difficulty convincing her of my case. Mrs. Ellul gave me similar accounts concerning some of the translations she checked. Second, philosophers of technology, almost without exception, have completely misread Ellul's work. I readily acknowledge that in translation Ellul's language may sometimes sound philosophical to English-speaking readers. However, Ellul was primarily a historian and sociologist.

One of the many sources of misunderstanding is the fact that the French language has no word precisely corresponding to the English word *technology*. The French language has two words: *technique*, which refers to the reality; and *technologie*, which refers to the discourse about that reality. As Ellul observes in Chapter 2, this is much as we distinguish in English between society and sociology. From its very beginning, the French were clear that industrialization involved not merely adding an industry and making a few accommodations as required, but completely restructuring human life in the world. In this sense the meaning of the French word *technique* is analogous to the meaning of the English word in statements such as, She is an excellent skier with a good technique, This pianist has an above-average technique, and so on. In the same vein, the French word is not in any way focusing exclusively on technology and industry. *Technique* refers to a different way of making sense of and dealing with the world than the one people had traditionally followed.

It may be helpful to contrast technique with culture, where the latter is interpreted as the totality of human creations that have sprung from symbolizing human life in the world. The following anecdote justifies this approach. As my five-year stay with Ellul was coming to an end in the spring of 1978, I presented him with a first draft of what was later published as *The Growth of Minds and Cultures*. After reading my manuscript, he asked me if he had not more or less said all of this already. I confessed that this was entirely possible, since I had not read everything he had written, and asked him to provide me with the appropriate references. He agreed, and after a few days we met again. He told me that nowhere, in fact, had he explicitly developed such a concept of culture. I readily acknowledged that this concept of culture was implicit in his courses and writings, however. I had developed it in part by asking how people got by before technique, and how this "before" continued to exist alongside technique in contemporary societies. Ellul concurred, and some of his other observations are set out in his introduction to my book. I will therefore attempt to clear up any confusion about the concept of technique by contrasting what I will refer to as the technical approach to life with the cultural approach.

Ellul defined technique as "the totality of methods rationally arrived at and having absolute efficiency (for a given stage of development) in every field of human activity."[2] This definition suggests a new, very different approach to human life and the world. The methods employed are no longer arrived at on the basis of experience and culture handed down from generation to generation as accumulated experience turned into a collective wisdom designed for living. The new methods are instead arrived at rationally, that is, by *ratio* (i.e., efficiency) rather than by context. Furthermore, these new methods are not concerned first and foremost with human needs, desires, or aspirations—or, for that matter, with human values of any kind. It is now a question of the methods having the greatest possible efficiency, calculated as the ratio of desired outputs and requisite inputs, without any reference to their meaning and value for human life. Individually, these methods limit the consideration of their immediate surroundings and the world to the inputs they deliver and the desired outputs they receive. Collectively, these methods restrict attention to a technical order built up from the technical processes that transform the inputs into outputs. These processes are connected by the exchanges of the inputs and outputs, including materials, energy, labour, capital, and knowledge, as well as products and services of all kinds. Such a technique-based connectedness is part of the fabric of relationships of a society and covers all spheres of human activities.

The past two centuries have involved a transition from societies relying primarily on the cultural approach to life, to mass societies relying primarily on the technical approach. The strengths and weaknesses of each approach reveal some of the profound differences in human life and the world before, during, and after this transition.

The Cultural Approach to Life

The cultural approach to life is based on the central role *logos* and culture have traditionally played in human life. *Homo logos* precedes *homo faber*. A tree branch must first be symbolized as something human before it can become a bow or a paddle. Plants and animals must first be symbolized as more than mere constituents of an ecosystem before they can be domesticated. Death must first be symbolized as something else before ritual burial becomes essential. Such instances of symbolization were part of a systematic attempt to make sense of and live in the world, by symbolizing immediate experience as moments of a life by relating them to all others and thus as moments of a human life lived within the group. Hence, every contact with the social and physical surroundings was symbolized relative to all others and thus as moments of a life lived in the world. In this way symbolization transformed our niche in the ecosystems of the biosphere into a symbolic universe. Naming everything in this symbolic universe signified the place and importance of everything relative to everything else in an individual human life—a life lived as a member of the group by means of a way of life in a world endowed with these meanings and values.

Thus, symbolization became an expression of both a human awareness of the world and also the influence the world had on that awareness in a particular time, place, and culture. Cultures and their ways of life embodied the experiences of many generations in a "project" or "design" for living.[3] What all these "projects" had in common was a knowing and doing based on situating everything in relation to everything else in human life by means of names, meanings, and values. Even the unknown was included. In this way, symbolization strengthened the ability of human beings to live a life and to make history, as opposed to merely participating in natural evolution. Each moment became integral to a person's life and that of his or her community, and each contact with an ultimately unknowable reality became integral to a liveable world. As a strategy for knowing and living in the world, the cultural approach to life is based on making the greatest possible use of context.

The importance of this can be seen by reflecting on the commonplace phrase "living a life." This connectedness of a human life is usually taken for granted, unless it is disrupted by a condition such as a short-term memory disorder or Alzheimer's disease. These conditions appear to interfere with the ability to symbolize new experiences by connecting them to the vast structure of neural connections built up in the brain in the course of living a life. The organization of the brain-mind symbolizes these experiences in ways about which little or nothing is known. New experiences can no longer be lived as moments of a life, instead becoming separate moments of existence, as it were. The separation of afflicted people's being—in time, space, and the social domain—cuts them off not only from their own lives but also from those of others and from their surroundings. For example, when such a person turns a corner in an unfamiliar building, he will be unable to retrace his steps to the front door because he will not remember the turn. Asking for directions is almost impossible because he will immediately be lost in any conversation. A person with such a condition no longer lives a life in a world, but inhabits a sequence of micro-worlds connected only by the life lived prior to the onset of the condition.

A similar loss of connectedness can occur at the level of collective human life. This is much more difficult to diagnose, and the findings will almost certainly be controversial. Nevertheless, the rise and fall of civilizations involve changes in the connectedness of collective human life. For example, a civilization risks collapse when it is no longer able to give meaning, direction, or purpose to the lives of its members, which can result in various groups going off in their own direction. The response of some ancient Greeks to such a situation has profoundly marked Western civilization. Socrates and Plato detected a weakening in the connectedness of Greek life because of its culture being relativized by contacts with others, and, in an attempt to establish a logical foundation for their culture, they sought to discover rational rules underlying daily-life experiences.[4] Much more recently, Karl Marx raised the question of alienation and the problem of a false consciousness as well as their distortion of the fabric of human life.[5] Emile Durkheim was concerned with what he perceived as a fundamental change weakening the connectedness of industrializing societies, and with the resulting anomie.[6] Max Weber observed the growth of rationality and its influence on human life, including what he called the disenchantment of the world.[7] Arnold Toynbee sought to explain the rise and fall of civilizations in terms of the constant need to adapt the connectedness of collective human life to new circumstances.[8] Ellul warned against the reification of human life under the influence of what he called technique.[9] Most recently, artificial intelligence researchers failed to describe the connectedness of daily-life experience in terms of the rules, algorithms, micro-worlds, scripts, frames, and other entities that were supposed to be its building blocks.[10] In the course of the twentieth century, our symbolic connection with reality, which earlier had often been taken for granted, became a subject for inquiry by the social sciences and humanities.

In the past, the cultural approach of any group or society included some science and technology that were unique to that approach and diffused along with it because they were integral to its symbolization of the world. Consequently, they were appropriate to the group's way of life and contributed to the ability of the local ecosystem to sustain that way of life. When, early in the twentieth century, science and technology became universal as components of technique, they broke with culture-based symbolization and helped create ways of life that became inadapted to a particular time, place, and culture as well as to a local ecosystem. The fact that concepts such as an appropriate technology and a sustainable way of life had to be invented in the second half of the twentieth century may be a sign of the cultural approach to life breaking down.

The Technical Approach to Life

The universalization of science and technology involved two interdependent developments. First, Western science and technology separated themselves from their host cultures, and consequently ways of life had to be adapted to science and technology instead of to what was cultural and local. Next, this spread to other spheres of life, becoming the cause and effect of the emergence of something much larger, namely technique, created by a technical approach to life. In contrast with the cultural approach, this technical approach represents a way of knowing and living in the world based on minimal use of context. Science, as the new objective and reliable approach to knowing ourselves and the world, does not tackle this task the way cultures do. This amounts to an anti-life and an anti-world orientation. Science behaves as if reality is unmanageably complex and cannot be known as such. Instead, most of the connections between what is to be studied and reality are severed by means of a process of abstraction. The object of study is thus placed in the manageable intellectual context of a particular scientific discipline and subspecialty, supplemented by the limited physical context of a laboratory experiment designed to examine a few variables, preferably one at a time.

Parcelling out the task of knowing ourselves and the world in this way turned out to be so manageable and efficient that science outdid its culture-based competitors. It is gradually beginning to dawn on us that in the absence of a science of the sciences capable of *scientifically* integrating their findings, each and every discipline separately contributes to an exponential growth of knowledge of things in a highly restricted context, as opposed to their "real-world" context. When such knowledge becomes more highly valued than its traditional counterparts, this simultaneously creates an exponential growth of ignorance of how things fit into, contribute to, depend on, and are inseparable from everything else. Discipline-based knowledge treats the findings of all other disciplines as knowledge externalities. In this way, each discipline establishes exclusive authority over a portion of the knowledge frontier, and a validation of any findings can be restricted to its practitioners. Transdisciplinary studies seek to reduce these knowledge externalities by placing a number of disciplines within a common context. However, the influence of such studies has been so small that disciplines continue to evolve within highly limited intellectual and physical contexts.

A similar approach for acting on the world by making minimal use of context emerged primarily within technology but quickly diffused to almost every other sphere of modern life, all but replacing the roles of tradition and culture. Specialists of all kinds are now engaged in adapting and evolving contemporary ways of life. Once again, these specialists all but sever the connections between what they are dealing with and the world, by means of a triple process of abstraction. First, they know the world beyond their domain of competence only in terms of the requisite "inputs" it delivers and the desired "outputs" to be returned to it. Second, a further abstraction is required because any domain of competence does not correspond to a "chunk" of the world. For example, no specialist knows everything about hospitals in terms of the process that takes in sick people, treats them, and discharges them back into the world. Doctors, nurses, pharmacists, technicians, social workers, nutritionists, office personnel, cleaning staff, volunteers, as well as relatives and friends all participate differently and partially in converting the "inputs" of sick people into "outputs" of patients on the mend. The more specialized the experts' knowledge, the more they deal with everything in terms of a "technical shadow" of its former self. Under extreme pressure, it may become "the appendix" in Room 1, "the gall bladder" in Room 2, and so on. What is coterminous with their domain of competence has been abstracted from the fabric of relationships associated with converting "inputs" into desired "outputs."

Third, decision alternatives cannot be adjudicated in terms of what is best for human life, society, or the biosphere because this would require that the domain of specialization be situated in relation to everything else. In the absence of this knowledge, what is "better" is operationalized in terms of how the technical shadow of a "chunk" of the world contributes to obtaining the greatest possible desired "outputs" from the requisite "inputs," or obtaining the same "outputs" from reduced "inputs." All other implications have been eliminated by this third abstraction. Success is measured in terms of output/input ratios including efficiency, productivity, profitability, cost-benefit ratios, risk-benefit ratios, and gross domestic product (GDP), which is obtained from relatively "fixed" inputs. Such performance ratios provide specialists with no guidance as to whether any improvement is partly or wholly achieved at the expense of the integrality of what is made "better," of the compatibility between what is made "better" and the broader context in which it operates, and of the ability of what is made "better" to evolve and adjust on the basis of self-regulation using negative feedback. It is obvious that the resulting "system" may accidentally get it right some of the time, but in most cases the "better" is achieved at the expense of context compatibility. Experts must shift their attention from the question, How can this improve our lives? to the question, How can this be made to yield its greatest power through converting requisite inputs to desired outputs?

To sum up, the technical approach begins by abstracting whatever is to be made "better," representing the remainder of the world only in terms of the "inputs" it must provide and the "outputs" it will receive. Next, whatever has been abstracted in this way is studied by further abstracting those features directly relevant to the goal of transforming the inputs into outputs, which are included in some kind of model, while the remaining features are excluded from it. The model is then manipulated to determine how whatever it represents can be made to function "better" in terms of contextless output/input ratios. The previously neglected contexts continue to be excluded. Finally, the results are used as a basis for reorganizing the portion of reality originally abstracted. The technical approach is a strategy for reorganizing human life and the world piece by piece, with minimal consideration given to the contexts within which these "pieces" occur. The strategy reifies each piece in the process. For example, when someone's work is reorganized on the basis of the technical approach, the work is not dealt with as an activity in the person's life. Similarly, modifying an organism by means of biotechnology takes no account of the integrality of the ecosystem in which the organism participates, thereby risking genetic pollution.[11] At bottom, the issue is a simple one. No one would think of tampering with a column integral to the structure of a tall building without understanding the function this column performs within the larger structure. Yet in biotechnology we are willing to tamper with the structure that supports all of life while having little or no knowledge of how particular genes or some other part of this structure helps support the whole. These parts are treated as objects for technical manipulation. The technical approach to life builds a new order. Insofar as it may be regarded as an order at all, it is an order of non-sense, because it is established and evolved outside of the domain of sense (i.e., the realm of culture). A distinction must be made here between nonsense and non-sense. Something is non-sense when it belongs to the domain of culture but violates its order of meanings and values. Something is non-sense, however, when it does not belong to the domain of culture even though it is a human creation. It creates an "order of disorder" within a culture, representing the equivalent of pollution within the cultural order.

Most of us could not imagine how we would do our work and get by without the technical approach. Indeed, contemporary ways of life rely extensively on such an approach and much less on the cultural approach. Self-evident as the technical approach may be, it excludes specialists from participating in the "system" as human beings. They neither know nor can take responsibility for the consequences of their decision-making because these fall mostly beyond their domains of competence. Nor can they participate, in the usual human sense, within these domains by placing everything in the full context of their lives and communities by means of experience and culture. Instead, they deal with everything on the basis of a knowing and doing that makes minimal use of context, thus separating themselves from experience and culture. They must also leave their (cultural) values behind.

Consequences of the Technical Approach

The creation of technique by a society involves a shift from a traditional society's almost exclusive reliance on the cultural approach to life, to a mass society's primary reliance on the technical approach to life (at the expense of the cultural approach). The strengths and weaknesses of each approach provide us with a key to understanding our recent journey. What are some of these strengths and weaknesses?

The replacement of experience and culture by the technical approach to life involves dealing with people and the world in a manner that externalizes everything not immediately relevant to the goal at hand. Nevertheless, throughout the entire process, whatever is abstracted, studied, modelled, simulated, and reorganized remains connected to everything left out. Consequently, the technical approach is very effective in obtaining the greatest possible desired results from whatever is required to produce them, but also in straining, distorting, or breaking many relations in the fabric of reality that were neglected, diminished, or marginalized by this approach. It also excludes the possibility of being guided by any human values, since the context of human life and the world is represented exclusively in terms of the desired outputs and the requisite inputs. It can be guided only by what masquerades as values but which is really nothing but output/input ratios. As a result, it disorders all that is shaped by the historical and natural processes through which everything in human life, society, and the biosphere evolves in relation to everything else. The technical approach to life is unmatched in getting results but does so at the expense of the integrality of the human and natural worlds. It creates a diversity of reorganized elements that in some respects are "carcinogenic" with respect to the cultural and natural orders.

As Ellul implies in his definition of technique, the technical approach to life cannot be applied in a piecemeal fashion. Since the very beginning of industrialization, it has been an all-or-nothing affair. Bettering one aspect of human life or the world in one place, and another aspect somewhere else, does more than locally introduce an element of chaos. Each technical improvement must be supplied with the requisite inputs and be relieved of the outputs it produces. Consequently, such improvements must be made in such a way that the outputs of one technical improvement constitute the inputs into another, linking all of them in a technical order that emerges by creating chaos in the cultural and natural orders. There can be no intermediary cultural or natural processes because these would not match the performance of the technically bettered areas. Hence, the technical order takes the form of a network of efficient transformations connected by exchanges of inputs and outputs. Where the technical order is connected to the cultural and natural orders, bottlenecks occur because the transformations within the network are more efficient than their predecessors and hence incompatible with the cultural and natural relations outside of it that remain untouched by the technical approach. These bottlenecks cry out for further technical improvements, compelling an ever-expanding technical order. Thus, the technical order must expand until it is coextensive with the cultural and natural orders. However, there is a constant tension between them as the technical order grows by undermining the other two.

In contrast, the culture-based approach to human life in the world is much less successful at obtaining desired results, but it respects as much as possible the way everything interacts with, depends on, and evolves along with everything else. After all, this approach is rooted in the way cultures symbolize any constituent of the world as being integral to human life, society, and the biosphere. In this sense, it is analogous to the way that all the constituents of any ecosystem have evolved in relation to one another over an extremely long period of time, thereby helping to constitute the varied conditions necessary to support this diversity of life. It is also analogous to the way that all the relationships that constitute a way of life evolve in the context of one another and the whole. For example, an invention is a response to a particular context, contributes to the evolution of this context, and will diffuse according to how well it does so.

This brief analysis of some of the strengths and weaknesses of the technical approach to life and its cultural counterpart can illuminate the successes and failures of contemporary ways of life and set them apart from all previous ones. We excel at improving the power and performance of cultural and natural relationships at the expense of the integrality and sustainability of human life, society, and the biosphere. Examples abound. The materials we create are extremely well-suited to the functions for which they are made but collectively undermine the health of ecosystems and, via them, human health.[12] We have surrounded ourselves with time-saving gadgets but have less and less time to ourselves.[13] The proliferation of communication technologies has been unable to overcome the "lonely crowd" or compensate for the kind of support individual persons once experienced from the kinds of social relations more common in a traditional society than in a mass society.[14] Computers are well-suited to a variety of tasks, but spending too much time in front of these machines negatively affects our image of what it is to be human and hence our relations with others.[15] Our means of transportation allow us to travel anywhere, only to find increasingly the "geography of nowhere."[16] High technology promised to deliver us from the smokestack industries and thus help solve our environmental problems, but instead it simply changed the kinds and quantities of pollutants being released.[17] The green revolution did not succeed in increasing the biomass obtained from photosynthesis, but instead boosted the edible portion of plants at the expense of the functions of the other portions, thereby necessitating more pesticides and herbicides with all the negative implications for soils and ecosystems.[18] Biotechnology, by not respecting the integrality of the DNA pool of the biosphere, is beginning to produce genetic pollution, with imponderable consequences. Our weapon systems are now so powerful that their all-out use can no longer defend anyone but only destroy everyone.[19] The information highway promised to deliver us from rush-hour traffic by enabling us to work at home, thereby reducing pollution levels in cities and making them more liveable. It would also provide more access to information, democratize the world, and do a great deal else. Instead, it has brought us more advertising, pornography, and loneliness. Pedagogical and educational innovations would help young people better adapt to the new realities, but instead they found it increasingly difficult to make sense of the lives of their parents, while parents intuitively realized that the new connectedness of the world had passed them by in ways that they could not grasp.

This pattern—of technologies that produce spectacular but specific results by undermining the cultural and natural orders—goes back nearly a hundred years. Human expectations tend to be based on these results without taking into account their "downside." The construction of the electrical grids was supposed to decentralize industrial production back to the home or small workshop, thereby eliminating many negative features of industrial society.[20] Nuclear power would feed these grids at a cost "too cheap to meter."[21] Workerless factories would shift the balance between work and leisure, to the point that we were supposed to worry about how we were to spend all our free time.[22] Educational television would bring the best teachers and professors within everyone's reach, thereby reducing the information gap between rich and poor, north and south, to create a more just world order. The microprocessor would revitalize democracy and decentralize large institutions. For a while, all this was celebrated as a move toward a more rational and secular society, until the "noise" caught up with the "signal" of technological and economic growth.

Our present situation is a continuation of these patterns. Engineers, managers, and regulators are accustomed to respectable figures indicating our success in obtaining desired outputs from requisite inputs as measured in terms of output/input ratios. However, when we regard the industrial-economic system as a whole, things appear very different. Alternative indicators to the GDP, which subtract costs incurred in the production of wealth from the total value of goods and services produced by an economy to arrive at *net* wealth production, show that the latter has been declining for decades.[23] The American Academy of Engineering estimates that 93 percent of materials extracted from the biosphere do not end up in saleable products.[24] Some time ago it was reported that Blue Cross was GM's primary supplier.[25] We can reconcile such contradictory impressions of how well modern ways of life serve us when we begin to interpret our successes in terms of signal-to-noise ratios of desired to undesired effects of design and decision-making. The former derive directly from what is abstracted from reality and included in the model, and the latter from what is externalized in the process. The former contribute to the successes of our economies, and the latter to their failures to prevent or minimize undesired consequences. To the extent that we pay for these failures, they directly undermine gross wealth production. In other words, the signal of technological and economic growth appears to be threatened by the noise of unwanted and unexpected effects. It would appear that the more we work on increasing the output of goods and services of modern economies, the more we are impoverishing ourselves. Economic development appears to be going in reverse.

The same kind of conclusion may be reached when we examine how industrially advanced societies deal with the unwanted effects of technological and economic growth. Contemporary societies are based on an intellectual and professional division of labour in which specialists of all kinds make decisions whose consequences fall mostly outside of their domains of competence, to be dealt with in an after-the-fact manner by others in whose specialties these undesired effects fall. Thus, the system first creates problems and then "solves" them. It is next to impossible to get to the root of any problem in order to prevent or minimize it. The system displaces rather than resolves the problems it creates, thereby feeding on its own mistakes and trapping us in a labyrinth of technology.[26] For example, we first produce pollutants, then install control devices to remove the most dangerous ones from waste streams, and then landfill them, which merely transfers these pollutants from one medium to another without solving the real problem. Also, we continue to feverishly restructure corporations to improve the productivity of labour, with the result, as shown by socio-epidemiology, that human work has become one of the primary sources of physical and mental illness.[27] The situation has compelled us to add social and health services at great expense.[28] Since these do not prevent the problem either, their costs can only grow to the detriment of corporations, employees, and society. All this raises the question of whether the primary outputs of technological and economic growth include waste and unhealthy workers. As these and other difficulties steadily increase, societies react by doing almost anything except going to the root of the problems. For example, we debate whether to privatize some or all health services, how best to increase the productivity of medical personnel and facilities by improving information systems, or how best to apply operations research to the scheduling of operating rooms. When this increases the stress levels of personnel even further, we add stress management clinics in an ongoing chain of compensating additions. Doctors order blood tests and other examinations, but many of them have little or no understanding of how workplaces affect us psycho-socially and physically. From socio-epidemiology we know that human health cannot be "produced" by means of disease care. Instead, it is sustained by meaningful and satisfying work, wholesome nutrition, adequate shelter, social support, and a fulfilled need to love and be loved. From this perspective, what we call the health care system is in effect an end-of-pipe disease care system, which has all but forgotten what health care is all about.[29]

To sum up: I have systematically developed the differences between the technical approach to life (which gives rise to mass societies evolving on the basis of technique) and the cultural approach (which gives rise to traditional societies and civilizations). By now it should be apparent that technology is but one branch of the much larger phenomenon of technique and that unlike technology, technique is co-extensive with culture.

The Autonomy of Technique

We know that each and every experience of our life is symbolized by the organization of our brain-mind, although we understand next to nothing about its higher symbolic functions. Now look around and examine your immediate surroundings. Apart from a few natural elements that have been incorporated into our habitat on our terms, everything is directly or indirectly the product of technique. Not only does this new habitat interpose itself between us and nature, but techniques also interpose themselves between us and other human beings: we use cell phones, watch television, or communicate by means of our computers. This mediation is not neutral, hence this new habitat of ours is likely to have as great an impact on the organization of our brain-minds and cultures as nature did in prehistory, and society during history. As people change technique, technique simultaneously changes people. If the influence of the latter is greater than that of the former, technique takes on a measure of autonomy with respect to human life.

In a nutshell, this amounts to a contemporary understanding of the human condition: whatever permeates our experiences will possess our brain-minds and our cultures. This is not some neo-Calvinist doctrine but a reflection of what we know from disciplines such as psychology, social psychology, sociology, cultural anthropology, and the sociology of religion. Nor is it a deterministic interpretation of human life in the world. There is no point in speaking about human freedom if there are no constraints; it is precisely these constraints that have always been the locus of struggles to free ourselves from what seeks to possess us. Ellul maintains that in our world the primary constraints are imposed by technique and the nation-state.

These alienate us, with the result that we are possessed by technique, much like the way capitalism possessed people in the nineteenth and early twentieth centuries. They also reify us, turning us and the world into objects for endless technical manipulation. A failure to understand this makes it impossible for us to exercise our humanity by engaging ourselves in a struggle against this alienation and reification. Here we encounter the point of departure for the other part of Ellul's work, which is the subject of the second appendix.

References

1. This appendix and the next are based on W. H. Vanderburg, *The Labyrinth of Technology* (Toronto: University of Toronto Press, 2000, 2002) and *Living in the Labyrinth of Technology* (Toronto: University of Toronto Press, forthcoming 2005).

2. Jacques Ellul, *The Technological Society*, trans. John Wilkinson (New York: Vintage Books, 1964), xxv.

3. W. H. Vanderburg, *The Growth of Minds and Cultures: A Unified Theory of the Structure of Human Experience* (Toronto: University of Toronto Press, 1985).

4. This point is most forcefully made by Eric Voegelin. See his *Plato* (Baton Rouge: Louisiana State University Press, 1966).

5. Bertell Ollman, *Alienation: Marx's Conception of Man in Capitalist Society* (Cambridge: Cambridge University Press, 1971).

6. Emile Durkheim, *Selected Writings*, ed. and trans. with an introduction by Anthony Giddens (London: Cambridge University Press, 1972).

7. Max Weber, "Science as a Vocation," in *From Max Weber: Essays in Sociology*, eds. H. H. Gerth and C. Wright Mills (New York: Oxford University Press, 1963), 129–56; Rogers Brubaker, *The Limits of Rationality: An Essay on the Social and Moral Thought of Max Weber* (London: Allen and Unwin, 1984).

8. Arnold Toynbee, *A Study of History*, abridgement of vols. 1–10 by D. C. Somervell (London: Oxford University Press, 1946).

9. Jacques Ellul, *The Technological Society*; Jacques Ellul, *The Technological System*, trans. Joachim Neugroschel (New York: Continuum, 1980); Jacques Ellul, *The Technological Bluff*, trans. Geoffrey W. Bromiley (Grand Rapids, MI: W. B. Eerdmans, 1990).

10. Hubert Dreyfus, *What Computers Still Can't Do: The Limits of Artificial Intelligence* (Cambridge, MA: MIT Press, 1992).

11. Miguel A. Altieri, "Genetically Engineered Crops: Separating the Myths from the Reality," *Bulletin of Science, Technology & Society* 21 (April 2001): 122–51; Miguel A. Altieri, *Genetic Engineering in Agriculture: The Myths, Environmental Risks and Alternatives* (Oakland, CA: Food First Books, 2001); Daniel Charles, *Lords of the Harvest: Biotech, Big Money and the Future of Food* (Cambridge, MA: Perseus, 2001); Brac de la Perrière,

Robert Ali, and Frank Seuret, *Brave New Seeds: The Threat of GM Crops to Farmers*, trans. M. Sovani and V. Rao (London: Zed Books, 2000).

12. Barry Commoner, "The Environmental Costs of Economic Growth," in *Energy, Economic Growth and the Environment*, ed. Sam Shurr (Baltimore: Johns Hopkins University Press, 1971), 30–65; Theo Colborn, Dianne Dumanoski, and John Peterson Myers, *Our Stolen Future: Are We Threatening Our Fertility, Intelligence and Survival? A Scientific Detective Story* (New York: Dutton, 1996); Deborah Cadbury, *The Feminization of Nature: Our Future at Risk* (London: Hamish Hamilton, 1997); David Weir and Mark Schapiro, *Pesticides and People in a Hungry World* (Oakland, CA: Food First Books, 1981).

13. Juliet B. Schor, *The Overworked American: The Unexpected Decline of Leisure* (New York: Basic Books, 1991); J. Gershuny, "Are We Running Out of Time?," *Futures* (January/February, 1992): 3-22; Benjamin Hunnicutt, *Work Without End: Abandoning Shorter Hours for the Right to Work* (Philadelphia: Temple University Press, 1988).

14. Hubert L. Dreyfus, *On the Internet* (London: Routledge, 2001); Scott Lash, *Critique of Information* (London: Sage, 2002); Laura Pappano, *The Connection Gap: Why Americans Feel So Alone* (New Brunswick, NJ: Rutgers University Press, 2001); Pippa Noris, *Digital Divide: Civic Engagement, Information Poverty and the Internet Worldwide* (New York: Cambridge University Press, 2001); Sherry Turkle, *Life on the Screen: Identity in the Age of the Internet* (New York: Simon & Schuster, 1995); Sherry Turkle, *The Second Self: Computers and the Human Spirit* (New York: Simon and Schuster, 1984); Craig Brod, *Technostress: The Human Cost of the Computer Revolution* (Reading, MA: Addison-Wesley, 1984); C. A. Bowers, *Let Them Eat Data: How Computers Affect Education, Cultural Diversity and the Prospects for Ecological Sustainability* (Athens, GA: University of Georgia Press, 2000); Orrin E. Klapp, *Overload and Boredom: Essays on the Quality of Life in the Information Society* (Westport, CT: Greenwood Press, 1986).

15. Sherry Turkle, *The Second Self.*

16. James Howard Kunstler, *The Geography of Nowhere: The Rise and Decline of America's Man-made Landscape* (New York: Simon & Schuster, 1993); James Howard Kunstler, *Home From Nowhere: Remaking Our Everyday World for the 21st Century* (New York: Simon & Schuster, 1998); James Howard Kunstler, *The City in Mind: Meditations on the Urban Condition* (New York: Free Press, 2001).

17. Faye Duchin, *The Future of the Environment: Ecological Economics and Technological Change* (New York: Oxford University Press, 1994); Michael Redclift, *Wasted: Counting the Cost of Global Consumption* (London: Earthscan, 1996). See also Barry Commoner, "The Environmental Costs of Economic Growth," 30–65, and Theo Colborn, Dianne Dumanoski, and John Peterson Myers, *Our Stolen Future.*

18. Miguel A. Altieri, *Genetic Engineering in Agriculture;* Brac de la Perrière, Robert Ali, and Frank Seuret, *Brave New Seeds.* See also Vandana Shiva,

Tomorrow's Biodiversity (London: Thames and Hudson, 2000); Vandana Shiva, *Monocultures of the Mind: Perspectives on Biodiveristy and Biotechnology* (London: Zed Books, 1993).

19. Robert Jay Lifton and Richard Falk, *Indefensible Weapons: The Political and Psychological Case Against Nuclearism* (Toronto: CBC Enterprises, 1982); Peter R. Beckman, *The Nuclear Predicament: Nuclear Weapons in the Cold War and Beyond* (Englewood Cliffs, NJ: Prentice Hall, 1992).

20. Neil Freeman, *The Politics of Power: Ontario Hydro and its Government 1906–95* (Toronto, University of Toronto Press, 1996); Jesse H. Ausubel and Cesare Marchetti, "Electron: Electrical Systems in Retrospect and Prospect," in *Technological Trajectories and the Human Environment*, eds. Jesse H. Ausubel and H. Dale Langford (Washington, D.C.: National Academy Press, 1997).

21. Neil Freeman, *The Politics of Power.*

22. Juliet B. Schor, *The Overworked American.*

23. Herman Daly and John B. Cobb Jr., *For the Common Good: Redirecting the Economy Toward Community, the Environment, and a Sustainable Future* (Boston: Beacon, 1989); Clifford Cobb, Ted Halstead, and Jonathan Rowe, "If the GDP Is Up, Why Is America Down?," *The Atlantic Monthly* (October 1995): 59–78.

24. Braden R. Allenby and Deanna J. Richards (eds.), *The Greening of Industrial Ecosystems* (Washington, D.C.: National Academy Press, 1994), Introduction.

25. Robert Karasek and Töres Theorell, *Healthy Work: Stress, Productivity, and the Reconstruction of Working Life* (New York: Basic Books, 1990), 11.

26. W. H. Vanderburg, *The Labyrinth of Technology* (Toronto: University of Toronto Press, 2000).

27. Robert Karasek and Töres Theorell, *Healthy Work.*

28. U.S. Department of Health and Human Services, *Mental Health: A Report of the Surgeon General* (Rockville, MD: 1999).

29. W. H. Vanderburg, *The Labyrinth of Technology.*

Appendix 2
Putting It All Together

JACQUES ELLUL'S WORK DIVIDES roughly into two parts: one describes what is happening to human life and society in the latter half of the twentieth century, and the other deals with the relevance Christianity could have in an age of science and technique. The following anecdote may trigger our intellectual curiosity about the nature of this division. After organizing a symposium to mark Ellul's death, I discovered to my astonishment that all but one of the speakers I had invited had become Christians through the reading of his work; and their understanding and beliefs, like Ellul's, had little in common with what often passes for Christianity today. Briefly put, for Ellul Christianity is an antireligion and an antimorality. His theology (if we can call it that) stands apart by being iconoclastic, and this fact hinges on a unique relationship between the two parts of his work.

Ellul wrote a book about an intervention in his life, a book which was not to be published until after his death. He reaffirmed this to me the last time we met. Apparently, he later decided that it should not be published at all, and it appears that he destroyed the manuscript. Given the influence and power of the Christian right, this action is not difficult to understand.

We can approach Ellul's theological work by first examining how we understand human life via cultural anthropology, depth psychology, and the sociology of religion. These disciplines reveal how today we are possessed by a secular sacred and myths, which already were the foundations for the great secular political religions that shook the twentieth century. This examination will facilitate our understanding of how Ellul interprets the entire Bible from Genesis to Revelation. He seeks to convince us that we are confronted with a new understanding of what it is to be human in an age of science and technique.

Each and every culture requires an ultimate reference point, which permeates daily life through a religion. In this there is little new under the sun; it makes no difference that today we do this in a secular way. Religion continues to make human life in the world possible, but it also continues to exact a high price. Failure to understand this key aspect of the human condition will inevitably turn Christianity into a religion and a morality. The Jewish prophets said it all before, although their message frequently fell on deaf ears because people were so possessed by the spirit of their age. Our present situation is similar because we too are committed to an absolute reference point and it also alienates and reifies us, with the difference that because of the power of our means, much more is at stake.

Homo Logos

In our scientifically, technically, and rationally oriented civilization, it has become very difficult to understand the meaning of the simple phrase "living a life." The relationship between each moment of my life and that life as a whole is somewhat analogous to the integrality of my physical self, where each cell in my body has enfolded within it the DNA that serves as a biological blueprint for the whole of my physical being. The analogy has some merit, because my DNA enfolds something of my parents, which in turn enfolds something of their parents, creating a tree-like structure of genetic relationships that ultimately links me to all of humanity and perhaps beyond to other life-forms. Is it also true that because my DNA includes a blueprint of my brain, and because this brain is in a small way modified by each and every experience, each cell also enfolds something of the interaction between natural evolution and cultural history? The validity of the analogy is further strengthened by the fact that my DNA can help produce new cells to replace old and worn-out ones, thus ensuring my physical continuity even though all my body cells (except for the brain cells) are replaced at least every seven years. The analogy can thus shed light on how my physical continuity enfolds the continuity of my whole being as a person of my time, place, and culture.

Can this analogy be extended to account for how the physical integrality of my being is connected to so much else in the world, past and present? How can I extend such an account to include the integrality of my whole being? In a time when so much of human knowing and doing is based on the scientific and technical approach to life and so much less on the cultural approach, attempting such an account runs against the mainstream in which all knowing and doing is based on minimal context. Despite the difficulty of giving such an account, it would appear that the invention of culture not only made us into "talking animals" but also supported the living of individual and collective human life in the world with as much integrity as possible.

The members of any society interpret their experiences and shape the relationships between themselves and the world into a coherent way of life by means of a culture. The considerable diversity of cultures can be interpreted as suggesting that the way human beings are linked to reality is genetically determined to only a limited degree. For a long time it was believed that the development of children was the result of the natural unfolding of universal states of mental and emotional growth. Although the roles of both nature and nurture were recognized, the learning of a culture was not regarded as the primary factor. Along with cultural anthropologists, I argue that by acquiring a culture, children learn from birth on to make sense of the world and to act in it in a way that is individually unique yet culturally typical. Culture-based symbolization thus acts as the basis for individual existence in reality.

Culture and Individual Life

Culture plays an important role in the way we maintain contact with the external world via the five senses. Consider some features discussed in my earlier study.[1] In an experiment designed to demonstrate this contact for the visual dimension of experience, subjects wore goggles that reversed the world from left to right and right to left. Initially this led to a great deal of confusion: talking with two friends involved hearing them in one place and seeing them in another; sitting down to dinner led to seeing a knife but feeling a fork and vice versa; and the scent of a passerby and the sound of her heels might come from the left while she was seen passing on the right. Eventually the brain-mind learned to reinterpret the visual dimension of experience to bring it in line with the other dimensions, even though the retinal images remained reversed. When, upon completion of the experiment, the subjects removed their goggles, the same kind of confusion arose once more, since the brain-mind had to again learn to reinterpret the visual experience in relation to the other dimensions.

From this and other experiments, it would appear that the interpretation of what is received from the optic nerve aims to determine the meaning for a person's life. Most of this interpretation is learned rather than innate. For example, newborns can follow movement but cannot focus their eyes until they learn that there is something to focus on. The visual interpretation by the organization of the brain-mind is further refined when toddlers learn to talk about their experiences, thereby aligning this interpretation with the way the language and culture of their community helps them to make sense of and live in the world. This is confirmed by significant cultural differences, as, for example, in the way the world of colour is organized, or by the way children who grew up in the wild made sense of their world. The visual experiences of adults are similarly affected by prior experience, as is evident when medical students learn to make sense of x-rays, which amounts to seeing something meaningful where before they might have seen meaningless blotches, experienced as a kind of "visual noise."

Our daily-life experiences confirm that what we see is not what is registered by our retinas, but instead what it means for our life. This becomes evident if we imagine an experiment in which what we see is compared with the output of a video camera mounted on a helmet (representing more or less what our retinas detect). Although the video output is framed, we do not see any edges around our field of vision. They correspond to nothing real and are therefore "interpreted out" by the process of symbolization. When we tilt our heads or go running, we symbolize the world as upright and stable, contrary to what is detected by our retinas.

Culture also plays an important role in the symbolic integration of the experiences derived from the external world in each of the five sensory dimensions, and from the body and the mind via several additional dimensions of experience. The foreground–background distinction of any experience depends on the way a person lives the corresponding moment of his or her life. Such experiences modify the organization of the brain-mind as they are said to be "stored" in "memory." The significance of this has been obscured in the English language by the use of the same word and concept for human and machine memory even though a great deal of experimental evidence suggests that these are fundamentally different. Machine memory stores information already separated from any context in a manner unaffected by any previous or subsequent storing of information. It is a contextless memory. Although machine memories can be used to simulate supposed neural networks, the context that can be incorporated in this way appears extremely limited compared to what the brain-mind is capable of. The latter permits the living of a life: each instance is symbolized as a moment of a life lived in a way that is individually unique yet typical of a time, place, and culture, by being symbolically mapped into the organization of the brain-mind. For humans it is not a question of mindlessly storing and retrieving facts, but of mindfully living a life in the world. Although almost nothing is known about the higher symbolic functions of the brain-mind, there is considerable evidence that the memory of an experience can be affected by earlier memories and can subsequently be affected by later ones. Human memory appears to make the fullest possible use of context. This characteristic of the brain-mind has evolved to cope with a living world in which nothing is ever repeated in quite the same way. Machine memory, in contrast, copes extremely well with the world of machines, in which everything is based on repetition and algorithms. If human memory had been genetically limited to functioning as machine memory does, humanity could not have survived.

In my theory of culture, I have described how much of human life can be explained if we adopt the admittedly simplistic analogy that the organization of the brain-mind "plots" each experience much as we plot the data from an experiment, in the sense that each experience may be regarded as a "data point" in our "experiment" of living in the world. Each experience is symbolically mapped in a structure of experience by synaptic and neural changes in the organization of the brain-mind. As we will see, this structure of experience may be compared to a "mental map" of a human life: a life relates to the organization of a brain-mind the way roads and cities relate to a road map. In a scientific experiment, each data point taken by itself provides us with very little meaningful information. This is still the case when a great many data points have been plotted, unless we are prepared to go beyond the data and to interpolate and extrapolate the evidence by fitting a curve through it. It is not until we have gone beyond the "facts" that meaningful new information about nature becomes apparent. If no curve can be fitted through the data, we will expect a flaw in the experimental design.

It is important to reflect for a moment on what this sort of behaviour means. What is the scientific basis for such curve-fitting? Have we not gone beyond the experimental evidence to leave the domain of science and enter the domain of speculation? Why does this strengthen our confidence in the data? The answer appears to be that fitting a curve through the experimental data confirms our nonscientific prejudgment of the world's behaviour as being continuous, nonrandom, and non-chaotic. Our confidence in the data is strengthened precisely because the curve confirms these prejudgments.

If we imagine internalization to be the "plotting" of an experience within the organization of the brain-mind, we might expect the development of a great deal of what I have referred to as metaconscious knowledge. It results from going beyond individual experiences to fully contextualize them in relation to all other experiences of a person's life. Much human behaviour confirms the development and extensive use of such metaconscious knowledge. Just as in a scientific experiment, what we learn about ourselves, society, and our physical surroundings is much more than the "data" taken one at a time would permit. The principal difference between living a life in the world and practising science is that in living a life we have no conscious access to how the brain-mind goes beyond our individual experiences and builds up metaconscious knowledge about our life in the world. Our prejudgment for doing so is an awareness that a healthy life is not chaotic, and that our surroundings at each moment are integral to our world.

Babies are born with a very limited ability to make visual sense of their world. Although they can respond to movement as a kind of reflex action, they learn to focus their eyes only when they begin to discover meaningful things to focus on. Their innate abilities are expanded by learning to "solve" visual "puzzles." Each visual experience appears to have the capability of functioning as a paradigm for others that are sufficiently similar but different. For example, the experience of seeing a cat may initially be little different from those of seeing other four-footed animals, and the specks of a bird and a plane flying high in the sky may initially appear more similar than different. Further visual exposure and the meshing with developments in other dimensions of experience cause the differences to grow, eventually overwhelming the similarities. This may be interpreted as a breakup of some visual paradigms into differentiated clusters of new ones, corresponding to the ability to make the distinction between dogs and cats or planes and birds. In this manner, each visual paradigm takes its place among all the others from which it has become differentiated, thereby permitting babies and children to live in a visually coherent world that gradually converges with that of adult members of their culture. Convergence is assured by the fact that at birth they are similarly embodied in the world, of which they become gradually more aware as they discover the meaning of participating in a body social. The visual development of children is, therefore, but one dimension of learning to make sense of and live in a world that evolves along with a growing awareness of their physical and social being. Each visual "skill" that babies and children acquire comes from the progressive differentiation of their ability to visually make sense of and live in the world, an ability from which this skill cannot be separated as a part of their "visual apparatus." I am not suggesting that paradigms, memories, differences, or similarities literally exist in the organization of the brain-mind, but that these are aspects of the process of symbolization.

The process of symbolizing the living of a life in the world cannot be piecemeal, mechanistic, or based on information processing, because this process of symbolization gets at the meaning and value of everything by placing it in the context of everything else in a life. All this is done from the vantage point or prejudgment of being embodied in the world as a human being. Elsewhere I have examined in some detail how this learning to make sense of and live in the world involves the development of a growing awareness of our physical embodiment in the world, our social selves as cultural beings, and a symbolic world derived from our experiences of reality.[2] Each step grows from an embryonic whole by means of a process of differentiation, somewhat analogous to the physical growth of an embryo by means of cell differentiation.

I have suggested that human memory is not a passive and context-less form of storage but an active part of the brain-mind, which organizes memories into larger patterns containing a great deal of metaconscious knowledge about how to relate to reality. Consider the example of the conversation distance that the members of a culture maintain, without being aware of it, when talking with each other. If we suppose that memories are directly differentiated from those that most resemble them, then the structure of the cluster of differentiated memories derived from these kinds of relationships will imply that if we stand too close to someone we are considered pushy, while if we stand too far away we are seen as unfriendly. The emotional tones associated with different parts of the structure will point to the culturally normal conversation distance. The structure of the cluster will imply a norm we learned without realizing it. The reason for this is that the knowledge we acquired is metaconscious, in the sense that it cannot be derived from any specific experience. It cannot be recalled from memory because it is generated by processes that systematically integrate internalized experiences on a level beyond that of consciousness.

In other words, a distinction must be made between the subconscious—repressed experiences as well as the knowledge implied in the genetically determined organization of the brain—and the metaconscious—the knowledge implied in the structure of experience constituting the mind. The latter plays a central role in the way that a culture structures individual and collective existence. It can be shown that in the course of being socialized into a culture, children build up metaconscious knowledge in their structures of experience. They implicitly learn such things as their culture's conversation distance, eye etiquette, conceptions of time, space, and matter, an image of their social self and the social selves of others, and the values and the way of life of their culture, including its myths and sacred.

The structures of experience of the members of a society may be likened to mental maps, with each individual's social self (metaconsciously derived from all their social experiences) as the map-reader, provided it is clearly understood that both the map and the map-reader are symbolically enfolded into each person's brain-mind. These mental maps permit people to orient themselves in their social and physical surroundings. In other words, the structures of experience of the members of a society form a symbolic medium through which they experience and act on reality. A great deal of our routine behaviour is modelled on typical earlier experiences. This, of course, does not mean that we are determined by our past. Our structures of experience include all aspects of our lives, including our hopes and fears for the future, our ambitions and plans, our dreams and fantasies, our convictions, thoughts, ideas, and so on; and at any time the routine usage of our mental maps can be overruled by thinking a situation through. A map differs from an algorithm or program because it requires someone to read and make use of it. This possibility is ensured by the metaconscious image of one's social self implied in a structure of experience. Since life never repeats itself, each paradigm contained in the map must be creatively adapted to the new situation.

As a result, much of the living of a human life can be coped with as a matter of routine, permitting a person to focus attention on those aspects that are particularly unusual, interesting, threatening, or for some other reason of particular meaning or value. For example, in a face-to-face conversation the eye etiquette, conversation distance, body language, and emotional expression require no special attention, thereby permitting people to concentrate on what is most essential. Again, even the most routine activity can be interrupted by a sudden thought that causes a person to rethink that activity, but all this helps to sustain the extraordinary complexity and diversity of human experience.

Culture and Society

We have thus far analyzed culture-based symbolization from the perspective of the individual. The characteristics of a society, however, cannot be derived from those of its individual members. If the links between an individual and reality are not genetically determined, neither are the links between a society and reality. In other words, a culture must not only symbolically mediate the relationships of the individual members of a society to reality, but also integrate the behaviour of individuals into a coherent way of life.

The members of any culture do not entirely know the reality in which they live. Modern science produces an ongoing flow of new discoveries about reality, and there is no reason to believe that this flow will ever come to an end. In other words, we must make a distinction between reality as it is lived by a society and reality itself. Yet, in their daily lives, the members of a society act as if reality as they know it is entirely reliable and differs from reality only in some nonessential details that remain to be discovered. At first sight this may not be surprising, because our intellectual heritage has told us for a long time that knowledge is cumulative and that we are, therefore, basically adding additional details to the essentially accurate gestalt of our knowledge of reality. This view, specifically for scientific knowledge, has been radically challenged by Thomas Kuhn, whose arguments concerning our knowledge of physical reality may be summarized as follows.[3] In the West, we have had very different ways of conceptualizing physical reality. There has been the Aristotelian view, which was followed by the Newtonian view, which in turn was succeeded by the Einsteinian view. Each of these three "pictures" of the physical world was elaborated during a period when that knowledge of reality was essentially cumulative. Such periods came to an end when it became apparent that the basic conception of physical reality was no longer adequate because a newly discovered phenomenon contradicted it. This contradiction ushered in a revolutionary noncumulative transition period.

Thus the growth of our knowledge of the physical aspect of reality cannot be regarded as a purely cumulative process. The basic gestalt of this knowledge changes from time to time; and in the absence of a complete knowledge of reality, it is impossible to say whether or not during a noncumulative period a more accurate picture emerges. All we can do is compare reality as it is known during different historical periods. During the cumulative periods, scientists behave as though reality is exactly as they know it except for missing details and improvements in accuracy. They speak of the laws of nature, for example, which are simply models that explain their experience of reality for a certain time. Subsequent generations of scientists typically discover that these earlier conceptions of reality embodied certain implicit assumptions and hypotheses that later on turned out to be incorrect. This is inevitable: scientists cannot but behave as if reality as they know it is reality itself, thereby implicitly assuming that the unknown has the same "nature" as the known.

The development of scientific knowledge within a particular discipline cannot be likened to what happens in an art class where students learn to draw a model. For an art student, the longer the pose the more time there is to add and refine details. Unless the gestalt is incorrect—such as when the student doesn't get the proportions right—the process is entirely cumulative, unlike the growth of scientific knowledge. Moreover, the "pictures" of the world drawn by different scientific disciplines cannot be integrated into a larger picture, because there is no science of the sciences.

If the growth of scientific knowledge is not cumulative, neither is the growth of culture-based knowledge, as is evident when we compare the knowledge that different or successive civilizations have acquired about the world. This situation raises some important questions. If we cannot assume that the unknown is simply more of the known yet to be discovered and lived, how reliable is the knowledge we already have? How do we know that some new discovery will not call our existing knowledge into question? How do we know that we can trust the world as we have come to know it? How do we know that we are sufficiently in touch with reality so as not to have to question our sanity? Somehow we must be able to trust our knowledge of the world, and this requires that the threat of the unknown be neutralized. It is one thing for philosophers to discuss, as an intellectual exercise, whether we are really here or whether the world is "out there," but it is impossible to live that way and remain mentally healthy. Imagine trying to do any activity such as walking, driving, or writing an exam if you had to worry all the time about whether what you were perceiving was really there. It is impossible to live with our human finitude without trusting reality as we know and live it.

All this points to the need a community has for a reference point to guide its journey in time, space, and the social realm. How this is accomplished follows directly from our earlier discussion of how the organization of the brain-mind symbolizes a structure of experience in which each experience of our life is connected symbolically to all others, thereby creating a great deal of metaconscious knowledge. Such knowledge interpolates and extrapolates the specific experiences of our lives in a manner that implies that the unknown has the same "nature." In fact, this metaconscious knowledge incorporates all specific experiences into a life lived in a world that now appears to be entirely seamless. The unknown is now metaconsciously symbolized as interpolations and extrapolations of the known, and it is only now that symbolization can reach its full depth. This completes the symbolization of each situation as a moment of a person's life, as an event in the collective life of a society, and as an integral part of the world of that society. It is the metaconscious equivalent of interpolating and extrapolating the experimental data in a scientific experiment to symbolize their full meaning by means of a curve.

The metaconscious interpolations and extrapolations of a person's experience correspond to what (in cultural anthropology, the sociology of religion, and depth psychology) are referred to as myths. Myths help gather individual experiences into a life, the lives of many people into a society, and the many contacts beyond that society into a world. It is this binding together that becomes institutionalized as a traditional or secular religion.

The practical implications of metaconscious myths are far-reaching. This becomes evident when we compare our daily-life dealings with reality with those of other human beings in earlier societies whose myths were "absolutely other" than ours. Indeed, that modern cultures also have myths appears to me to be an inescapable fact, because modern science and technology cannot fundamentally alter the condition of human finitude in an ultimately unknowable reality. Our knowledge also must be grounded in hidden metaconscious interpolations and extrapolations that amount to hypotheses and assumptions about the nature of reality and our existence within it, which correspond to the myths of a society.[4] Although we know how the ways of life of societies in the past were grounded in myths, we are generally unaware of the myths that underlie our own existence. These will undoubtedly become apparent to future generations, but in the meantime we act as if our lived reality is reality itself. This implies the elimination of the relative character of our life's knowledge by *absolutizing* reality as we know it. It also implies that our culture symbolically dominates reality. In other words, what is unknown and therefore threatening to the stability of our knowledge and our lives is converted into mere extrapolations and interpolations of reality as we know and live it. By absolutizing reality as it is known by a society, a system of myths converts the unknown into missing bits and pieces of the known. It helps convert *a* way of life into *the* way of life by making all alternatives unthinkable and unlivable.

To sum up, the metaconscious processes that integrate the experiences of a life into a coherent whole close the gap between reality and reality as it is lived, effectively obscuring from consciousness all alternate possibilities of interpreting and living in reality. The unknown becomes simply a storehouse of missing details to be added to the reality as it is known and lived. The system of myths of a society is therefore an important element in the creation of its cultural unity, because a different awareness of human life in the world becomes existentially impossible. Myths, after all, point to what reality will almost certainly be like based on all available experience. They extrapolate and interpolate between all available experiences to create a coherent picture of our lives, our society, and the universe, otherwise our mental maps would simply be a set of incoherent and only loosely related fragments. The metaphor of connecting individual experiences into metaconscious patterns can help us understand how important myths are to human life in the world. Each experience, symbolized as an alteration in the organization of the brain-mind, becomes a moment of the larger "pattern" that symbolizes a person's life. Experiences of encounters with others are symbolized as moments of their lives, and experiences of the physical surroundings become those of a meaningful world. In this way, culture sustains and reinforces the living of individual and collective human life. Distinct and separate contacts with others and the world are thus transformed into a meaningful, purposeful, and livable world. Myths are not merely the connections between these contacts but the very life that makes these contacts possible.

Our contemporary understanding of myths is the exact opposite of how they were understood in the nineteenth century—namely, as the religious and superstitious remnants of a distant past for which there would soon be no place or role in individual and collective human life. Myths still act as a "spiritual force" that first and foremost binds people together by means of a common reference point. It comes at the cost of alienating individual and collective human life by possessing it the way a master possessed a slave in earlier societies. In industrial societies it is the content of myths rather than their role that has changed, a fact that becomes evident when we examine how the religious dimension of these societies became secularized to produce new political secular religions including communism, national socialism, and hard-line democracy. Myths continue to be the meta-conscious roots of all (secular) religious expression. In the same way as there can be no science without going beyond the experimental data, there can be no human life in a "cultural niche" without myths.

On the deepest level, we might say that a society's system of myths acts as a kind of cultural DNA. Thus, each experience becomes a moment of our life, just as each body cell enfolds something of our biological whole. Similarly, everything in our external world is transformed into an element of the symbolic universe of our culture. Since the meaning and value of anything in that universe is an expression of its place and significance relative to everything else, it enfolds something of the whole and evolves as such.

A system of myths thus reduces the threat of the unknown and eliminates the relativity of a society's knowledge of reality. Yet, this is still not enough to provide societies and civilizations with the kind of stability and endurance seen in history. All human activities made possible by the absolutized lived reality would be equally good or bad, equally useful or useless, equally beautiful or ugly, and so on. In other words, without the system of myths performing additional functions, each moment of our life would be equally meaningful, that is, without any meaning at all. Choices might as well be made randomly, because there would be no possibility of meaning. Life would be a random sequence of events, a complete chaos which would be existentially unbearable. This is true for both individual and collective existence. The members of a society must be oriented in the "space" created by all possible relations and be shown how to act in it. Every culture achieves this by means of a hierarchy of values anchored in its system of myths. The latter also gives the members of a culture a reason for living and motivates them to adopt a way of life that has meaning and value for them. Reality as it is known and lived must become a society's home—what we have called its symbolic universe.

The values of a society reflect the basic vitality of a culture. Generally speaking, the metaconscious structure of experience tends to identify one or more phenomena in the life of a community that so permeate it that its very existence, and thus also the lives of its members, becomes inconceivable without them. For the prehistoric group, such a phenomenon was what we would call nature, and for the societies that began to emerge at the dawn of history, it became society itself. In other words, these phenomena correspond to the primary and secondary life-milieus for human life.

The metaconscious recognition of these kinds of phenomena confronts a community with a dilemma. The community could decide that such a phenomenon is so all-determining that it has little or no control in the face of this fate. On the other hand, and this is in fact what happens, the community could secularize the phenomenon by metaconsciously bestowing an ultimate value upon it. Necessity is thereby transformed into "the good," and the social order is the expression of the community's members freely striving for that good. The freedom and cultural vitality thus metaconsciously created eventually permit the sacred to be transcended as an all-determining force and to make human history possible, although exacting a heavy price. All this may be put into traditional religious language when we recognize that such phenomena are so central and fundamental to individual and collective human life that it would be unthinkable and unlivable without them. In this sense, they have created that life and the world in which it is lived. They become the creators and sustainers of life and its absolute moral authority. As we will see shortly, this metaconscious religious operation has nothing to do with the possibility of there being a transcendent. At the dawn of history the overwhelming influence of what we would call nature was slowly transcended, although that of society eventually took its place. In turn, this societal influence was eventually replaced by that of technique.

The bestowing of an ultimate value upon whatever is most central and determining in the life of a community metaconsciously orders all other values in the structures of experience of its members. People live as if what is most important in their lives is "the good." Nothing more valuable, important, and life-sustaining can be lived or imagined. This is why the sacred is also the central myth. Thus, the absolutization of reality as it is known and lived creates a sacred, a system of myths, and a hierarchy of values, which together constitute a cultural unity. Because it is profoundly metaconscious, this unity gives a great deal of stability to a culture as well as giving it a history distinct from natural evolution.

The cultural unity embedded in the structure of experience of every member helps structure thought, communication, and social behaviour. This unity makes the social order self-evident and natural. It provides the basic models for responding to new situations. Myths are not directly experienced by the members of a culture; rather, it is through myths that the world is experienced. The unity of a culture is a symbolic equivalent of DNA.

When the members of a society intuit what their metaconscious has identified as the sacred—the phenomenon that is attributed the highest value—they do not treat it casually but as something that is very special; that is, they tend to treat it with religious awe, as the value of values. Life without this most valuable entity would be unimaginable, unlivable, and unbearable. Who would they be? What would their life be like? What world would they live in? Symbolically, the sacred has made them and their world who and what they are. To put it in religious language, this sacred is the creator and sustainer of themselves and their world. To put them in contact with this sacred in order to influence it, a culture's religion is developed around this sacred. Without a cultural unity the members of a society would not have firm roots in reality and no order or meaning for their lives. This "reaching for the heavens" reassures them that they are really in touch with the universe and that their lives are meaningful, which is necessary since their relationships with reality are not sufficiently determined by means of innate structures of the brain, as is the case for animals. The establishment of a cultural unity in the metaconscious patterns of people's structures of experience ensures that each moment of their lives is lived in the context of this unity; it permeates all experience.

No longer do the members of a society have to be preoccupied with the ultimate questions. These have been put to rest by the metaconscious creation of a symbolic cultural unity. By working out the relationship with the sacred by means of a religion, a society reassures itself that it is not lost in reality, that the universe is no longer something it does not understand and over which it has no control. Hence, life and death become bearable. A society can put itself in contact with the powers of its world by personifying its sacred. The future continues the present, since anything that is "wholly other" is unthinkable and unlivable. All these and many other functions of religion are well known. We simply need to stress here that we are not debating the existence of a transcendent God who reveals Himself to, and communicates with, human beings. This is a separate matter that cannot be ruled out by the theories of religion as a cultural element, as even Karl Marx implicitly recognized.

Revelation and Religion

From conversations I had with Ellul during and after my stay with him, I believe that the above interpretation of human life and the world roughly corresponds to his understanding prior to an intervention in his life. At the time, he was thoroughly familiar with Marx's interpretation of religious phenomena as a consequence of a false consciousness. The intervention was the act of a God who was not the religious fabrication of his culture but who was holy, which simply means being separate from Ellul's cultural world and its religious fabrications. Such a God could not be fitted on Ellul's "mental map," even though the experiences of that encounter were symbolized like all others. This symbolization revealed only their cultural and religious meanings and values because they enfolded his cultural unity. He decided to think through these experiences in the context of other encounters between this God and people, which had been handed down from generation to generation and eventually recorded and accepted as the Jewish and Christian Bibles. Ellul recognized that these records were not a "pure" Revelation, since the experiences of these people were also symbolically mediated. Hence, these texts could become Revelation only when the meaning and value were, in a manner of speaking, lifted off the pages by further Guidance. This too was symbolically mediated.

His encounter with God placed Ellul in a very difficult situation. To make an intellectual distinction between Revelation and religion is one thing, but to live it is quite another. It is impossible to live without one's culture for making sense of and living in the world, but this culture implies a defining commitment that makes all of us people of our time, place, and culture. It is equally impossible to substitute the Revelation for one's culture. Ellul had become a witness to an event that was *in* his life and culture but not *of* them.

For Jews and Christians, the sacred and the religion based on it that all cultures have created must be distinguished from the "Wholly Other," who is not a cultural creation and whose communication with people enters into the symbolic universe of a culture as a Revelation (which must be distinguished from religion).

Although Ellul accepts this basic distinction between religion and Revelation, he goes a step further by recognizing that in the lives of Jews and Christians the two inevitably mingle together. By being socialized into a culture, Jewish and Christian believers acquire a cultural unity like anyone else, and this cultural unity includes a sacred. In other words, to belong to a society is to violate the first of the ten commandments. Again, it is as if the reality as it is known and lived by a community is reality itself, and as if the greatest good known by a community is the good itself. Making a commitment to the Revelation does not make the culture of the believers go away. In their lives are sown both wheat and tares, and one cannot be uprooted without the other.

In addition, the Revelation as passed on in oral or written form requires interpretation, and this activity can be carried out only by means of the culture of the believer. The experience is somewhat like wearing "cultural glasses," except that a person cannot take them off. This becomes a serious problem when the interpretation of the Revelation is influenced by theologians who are not iconoclastic with respect to their culture. Since believers are people of a time, place, and culture, they must be iconoclastic with respect to the absolutes created by their culture, and this iconoclasm must influence how they read, interpret, and practise the Revelation. With respect to contemporary science and technique, most recent theologians have been particularly lacking in an iconoclastic attitude. This has created endless distortions. For example, it became fashionable to read the first few chapters of Genesis as a flawed scientific account. The question should have been asked as to why, precisely when the theory of evolution was beginning to have an influence, was it then that believers began to read these first few chapters differently. Was this not a sure sign that things were being read into these texts that earlier generations did not see? Examples of this kind can be multiplied as the Biblical text was subjected to one scientific fashion after another. Ellul points out that today technique and the nation-state constitute the new sacred, and that science and history are the principal myths. Much theology in the twentieth century used the cultural absolutes or near-absolutes of technique and science to judge the Revelation, thus making complete non-sense out of it. Protestant and Catholic churches alike, as institutions attempting desperately to be relevant, adopted one technique after another, but more importantly, adapted the Revelation to the reference points of contemporary cultures. Theology, evolving much like any other discipline within the contemporary university, is almost without exception non-iconoclastic with regard to the reference points by which we live and journey on our way.

By now it should be apparent that it was next to impossible for the Jewish people to avoid synthesizing their need for a culture (including its religion) with the Revelation. The prophets constantly had to draw them back with dire warnings of the consequences. Christians faced the same pressures, with the result that, within a very short time after the life of Jesus, Christianity began to serve a religious function, first in the Roman Empire and later during the Middle Ages. Still later, Protestants synthesized the Revelation with capitalism. Protestants and Catholics alike greatly assisted this system in extending its ecological footprint by means of colonization. It is all too often that Catholic and Protestant churches have succumbed to the religious pressures of their cultures. One of the most recent instances was seen during the U.S.-led war against Iraq, when conservative Christian Americans declared on the media that God was right behind their president.

In my classes and public lectures I have frequently tried to make people aware of the presence of myths in our own society by means of the following exercise. Make a list of five items that are inaccessible to science. Next, make a list of five items that cannot be improved by technique. Finally, create another list of issues that are nonpolitical. When people have difficulty making such lists, it becomes evident that their creativity and imagination are limited by their metaconsciously regarding science as omnipotent in the domain of knowledge, technique in the domain of action, and the nation-state in public life. We have enormous difficulty seeing science and technique as human creations that are good for certain things, harmful for others, and irrelevant to still others. If we do not see them in this way, we have little chance of making much sense of the Jewish and Christian Revelation. I believe this is the key to understanding Ellul's sociological and historical work in relation to his theological work. It is also the reason why he was in constant tension with institutionalized Christianity.

Making sense of human life—from the difficulties faced by youth to the message of modern art—Ellul, like everyone else, made use of his culture. Since this use implies a commitment to a reference point, doing so inevitably distorts the integrity of anything that interferes with it. To deal with this key aspect of the human condition requires an iconoclastic attitude with respect to our commitments, and this in turn requires another reference point. In the life of Jacques Ellul this established a tension between, on the one hand, our contemporary sacred, myths, and secular religions and, on the other, his commitment to a God who unconditionally loves His creatures. This tension is the conflict between the Revelation of that love and our secular religious commitments to technique and a nation-state. It is the latest form of the profound struggle between Revelation and religion that has been an integral part of Western civilization from its beginning, accounting for much of its dynamism. The adaptation of Christianity to a personal religion has provided us with the well-known heart, consolation, and hope in an age without heart, consolation, or hope. The West is losing its dynamism precisely when its most influential creations—namely, science and technique—are spreading across the rest of the globe.

It is now possible to shed more light on the relationship between Ellul's sociological and historical writings and his "theological" ones. As a sociologist and historian, he attempted to understand human life, with the knowledge that his cultural vantage point and "mental lens" implied a defining commitment different from the one he sought to live. As a theologian, he also attempted to understand human life in the light of Revelation, knowing full well that his reading of the Jewish and Christian Bibles involved the same cultural vantage point and mental lens. To the extent that he could live by faith and accept that he was loved, he was able to challenge the cultural ground underneath his feet. Living this way involved calling into question everything that appeared certain and trustworthy in his life without falling prey to anomie, depression, mental illness, or even suicide. Iconoclasm is possible only to the extent that one is able to give up the religious assurances of one's culture. Faith can only come at the cost of doubt. Hence, Ellul's parallel attempts at making sense of human life and the world constantly challenged one another, and this in turn affected his cultural vantage point and mental lens.

Here we begin to understand why the Christian Bible insists on believers becoming like children. The human condition is not the result of a wrong turn in the evolution of the human species. The brain does not "wire" human life for alienation. This becomes evident when we compare the behaviour of children with that of adults. Children behave as if they know that the reality as they have come to know it is not reality itself. They accept that their mental maps present them with a world of which they have inadequate knowledge. For example, children may confidently stride into a big store at the hand of a parent, but if they break away to go and explore and then become lost, they quickly become anxious and may even burst into tears. No longer having access to a parent to show them how to live in the world, and to rescue them if they are stuck, calls everything into question. In other words, children do not act as if their mental map is complete and reliable; they are open to outside guidance. Adults, on the other hand, choose to live as if their mental map is the final word on everything, and this makes it impossible for them to live any longer in the world with a certain playfulness as children do. In other words, adults live as if their structures of experience are algorithms, thus robbing themselves of the freedom their mental maps might bring. If we struggle against being possessed by our cultural unity (the source of our alienation), we must engage in a lifelong struggle that is iconoclastic with respect to our culture. Our structures of experience are nothing more than an expression of our human finitude, and there is nothing absolute about them.

In the Jewish tradition, the danger of creating gods was paramount. The First Commandment is directed at this issue. According to the Torah, after the break between God and humanity and the banishment from the Garden of Eden, the need to collectively make sense of and live in the world necessitated the building of the Tower of Babel. The situation was somewhat analogous to the one we described about toddlers going shopping with a parent. The Tower of Babel symbolized reaching toward the domain of the (culturally created) gods in order to be in touch with reality once again. Without the Tower, there could be no certainty whatsoever and no firm ground to stand on, which would make human life unbearable and impossible. It would appear, therefore, that the Tower of Babel symbolized, not the creation of cultures, but the substitution of a cultural unity (with a sacred, myths, and a hierarchy of values) for God Himself. The Tower restored an essential certainty and ground for human life, but at the price of sin. It must be emphasized that sin was compared to the condition of slavery, which makes it impossible to be truly human because being possessed by a master and thus not being free makes any relationship of love for God and neighbour impossible. Who would take seriously a declaration of love made by someone under the threat of the whip?

What all this implied was largely lost to Western civilization until a secular equivalent understanding appeared with Ludwig Feuerbach, and more fully with Marx. They recognized the fundamental importance of religion for human life and also that its benefits came at the terrible cost of alienation—the secular equivalent of sin. The Christian community produced very few iconoclastic theologians. (Ellul told me that, following the publication of his book *The Presence of the Kingdom*, one of the greatest theologians of the twentieth century wrote him to inquire what the meaning of technique was.) I believe that as a theologian Ellul stands out for his iconoclasm, which is expressed through the ongoing dialogue between the two parts of his work. For example, his studies of phenomena that can alienate human life include the sociological analysis of technique as well as the theological analyses of the city and the role of money. The role of Revelation in human life is theologically examined as the presence of the Kingdom, and sociologically analyzed as a false presence of the Kingdom. The sociological analysis, in *The Political Illusion*, of politics in contemporary life has its theological counterpart in *The Politics of God and the Politics of Man*. Many other subjects are examined in this dual way. Unfortunately, much of the intellectual gift Ellul has left us has been received either as exclusively theological or as sociological and historical. Few people understand the essential relationship between the two approaches, which sustains Ellul's iconoclasm and struggle for freedom in response to a revealed love.

All this is somewhat reminiscent of the accounts some people have given of their near-death experiences. Bathed in a reassuring light, they "re-lived" their lives, from which the veil of their commitment to a sacred and myths had been lifted. It was a terrible thing to understand what they had really done to others and themselves; and without that light it would have brought on an ultimate despair. In the same vein, to the extent that revealed love gives us the courage to be iconoclastic with respect to our religious certainties and lift the veil from our lives to face who we really are, we can also better understand the reality of our world. To the extent that we can do this in the faith of the ultimate reality of being loved, to that extent will we be able to do without the religious assurances of our cultural unity. However, this makes us sojourners without a cultural home. In the life of Jacques Ellul, an escape from being possessed by the many scientific fashions that articulated the cultural unity of his society permitted him to discern where we were going, with greater clarity than any other social scientist that I know of in the second half of the twentieth century. Few heard his warning, and even fewer understood the source.

References

1. W. H. Vanderburg, *The Growth of Minds and Cultures* (Toronto: University of Toronto Press, 1985).
2. Ibid.
3. Thomas S. Kuhn, *The Structure of Scientific Revolutions*, 2nd ed. (Chicago: University of Chicago Press, 1970).
4. See, for example: Roger Caillois, *Man and the Sacred*, trans. M. Barash (New York: Free Press, 1959); Mircea Eliade, *The Sacred and the Profane*, trans. Willard R. Trask (New York: Harper and Row, 1961); Mircea Eliade, *Patterns in Comparative Religion*, trans. Rosemary Sheed (Cleveland: World, 1970); Jacques Ellul, *The New Demons*, trans. C. Edward Hopkin (New York: Seabury Press, 1975); Claude Levi-Strauss, *The Raw and the Cooked*, trans. John and Doreen Weightman (New York: Harper and Row, 1969); Richard Stivers, *Evil in Modern Myth and Ritual* (Georgia: University of Georgia Press, 1982); Paul Ricoeur, *The Symbolism of Evil*, trans. Emerson Buchanan (New York: Harper and Row, 1967); W. H. Vanderburg, *The Growth of Minds and Cultures*.

Anansi Nonfiction

Ivan Illich in Conversation
by David Cayley

Polymath and polemicist Ivan Illich alights on such topics as education, history, language, politics, and the Church. These conversations range over the whole of Illich's published work and public career, and illuminate his importance as one of our era's great thinkers.

0-88784-524x

Finding Peace
by Jean Vanier

Jean Vanier reflects on world events, such as the terrorist attacks in the U.S. on September 11, 2001, identifying the sources of conflict and fear within and among individuals, communities, and nations that thwart us in our quest for peace. Vanier shows us that ordinary people, unknown and unrecognized, are transforming our world little by little, finding peace in our neighbourhoods and lighting the way to change.

0-88784-6831

The Real World of Technology (Revised Edition):
The CBC Massey Lectures Series
by Ursula M. Franklin

Renowned scientist and humanitarian Ursula M. Franklin examines the impact of technology on our lives, tackling such issues as the effects of virtual space and collapsed time on our communities, the impact of communications technology on government and governance, the shift from consumer capitalism to investment capitalism, and the influence of the Internet on education and the sharing of knowledge.

0-88784-636x

The Cult of Efficiency (Revised Edition):
The CBC Massey Lectures Series
by Janice Gross Stein

Janice Gross Stein reveals how the discussion of efficiency in education, health care, and security can often be a cloak for political agendas, and calls for greater accountability and choice in the delivery of these public goods.

0-88784-6785

Available at fine bookstores or at www.anansi.ca